D0203219

THE MAMMALIAN CELL AS A MICROORGANISM

Genetic and Biochemical Studies *in Vitro*

THE MAMMALIAN CELL AS A MICROORGANISM

Genetic and Biochemical Studies *in Vitro*

Theodore T. Puck

Eleanor Roosevelt Institute for Cancer Research
Department of Biophysics and Genetics
University of Colorado Medical Center
Denver, Colorado

hd

HOLDEN-DAY, INC.
San Francisco
Cambridge
London
Amsterdam

THE MAMMALIAN CELL AS A MICROORGANISM
Genetic and Biochemical Studies *in Vitro*

500 Sansome Street, San Francisco, California 94111

Library of Congress Catalog Card Number: 73-188127
ISBN: 0-8162-6980-7

1234567890 MP 798765432

Printed in the United States of America

PREFACE

In the middle of the twentieth century, a revolution occurred in biology which changed forever the fundamental character of this science. These developments, which acquired the name of molecular biology, have provided life with a new theoretical basis in which living behavior has become firmly based on the behavior of specific molecules. While many of the properties of living organisms still remain to be elucidated on this basis, molecular biology has explained so many fundamental biological properties in terms of the basic concepts of physics and chemistry that there seems little doubt that further extension of these principles will lead to still deeper and broader understanding of biological behavior. Moreover, this new approach has demonstrated so many basic molecular events that are similar in all living cells, that the conclusion is inescapable that life processes are now more fundamentally understood than was ever before possible.

Molecular biology arose primarily out of studies carried out with the simplest living organisms: molds, bacteria, and viruses. In the early twentieth century, biology, expanding in many directions, had demonstrated the universality of the Mendelian genetic laws and the key role played by the enzymes in life processes. While many cellular structures were cataloged and cell functions defined, the various fields of biology remained isolated with relatively little interaction. For example, developments in biochemistry had demonstrated that proteins consist of chains of amino acids, and that the same twenty amino acids are utilized by all living organisms; it had established the catalytic role and specificity of enzymes whose mediation of the complex chemistry of the cell makes life possible; and it had delineated many of the characteristics of hundreds of separate enzymatic reactions which are carried out by living cells.

It is no wonder then that the biochemists came to feel that the ultimate secret of the living process must lie in the proteins, and particularly in the enzymes. However, equally important developments were taking place in genetics. Mendel and his followers had demonstrated that the hereditary characteristics transmitted by all living things to their offspring are contained within certain unit characters present in every cell. These genes, as they came to be called, had not been identified as material particles, but could only be inferred as abstractions, whose existence was postulated from the fact that the offspring of particular matings yielded progeny of different kinds, whose frequencies approached simple ratios. However, since presumably every characteristic of every organism depends upon such a specific unitary element, it seemed natural for the geneticists to conclude that the ultimate secret of the living process lay in the nature of the gene.

The great synthesis that paved the way for the creation of the new molecular biology developed out of the work of Beadle and Tatum, who announced the "one-gene-one-enzyme" hypothesis. By simultaneous biochemical and genetic studies on the cells of the simple bread mold, Neurospora, they were able to show a one-to-one correspondence between particular genes and particular enzymes. Thus the apparent dichotomy between biochemistry and genetics was bridged. This conceptual synthesis was followed by an explosive series of developments in which among other things the gene was shown to consist of deoxyribonucleic acid (DNA) molecules; its three-dimensional structure was deduced; its mode of replication was elucidated; the nature of the information storage by which each gene specifies the amino acid sequence of the protein chain for which it is responsible was demonstrated; and the mechanism of protein synthesis, by which the blueprint contained in the gene is translated into a specific protein structure, was at least in large outline developed. These episodes, which constitute one of the most dramatic sequences of man's intellectual history, brought about an enormous unification of many of the highly diversified activities which had, until then, constituted the science of biology. A brief resume of the concepts of molecular biology is presented in Appendix I.

Like *E. coli*, the mammalian organism begins life as a single cell which is capable of reproduction. In addition, however, the fertilized egg initiates an orderly sequence of events known as cell differentiation. These steps involve a progressive turning on of certain gene sequences and turning off of others in a fashion which appears to be rarely or never reversible under natural conditions.

The complexity of the mammalian organism is therefore greater than that of *E. coli* at several different levels. The size of most mammalian cells is roughly 500-1000 times greater, and some new structures are made necessary by this increase alone. In addition, new structures are required for each cell to assume its own specific differentiation pathway. Finally, the mammalian genome contains approximately 500-1000 times more DNA so that its genetic potentialities are greatly increased. However, these differences alone do not furnish a measure of the true difference in complexity between *E. coli* and the complete mammalian organism. The mammal is an organism in which approximately 10^{13}-10^{14} cells are united to form a new complete whole. The functions of the whole organism, however, are only possible in terms of the contributions of each individual cell, and these cellular associations result in an enormously increased range of powers as compared to an organism like *E. coli* in which each cell is complete in itself. Consider as an analogy, a comparison between a single relay, which is capable only of on-off responses to an input signal, and a computer, which contains millions of relays linked together in a highly organized fashion. Such a computer is capable of executing complex operations of a kind inconceivable if one is limited to any number of single independent relays. Thus mammalian behavior entails greater complexity than that of bacteria not only in the number of individual cellular units but also in the ability of such units to become coupled in vast arrays of highly specific organizational patterns. Understanding of the mammalian organism requires study at all of these levels.

The purpose of this book is to describe some of the studies on somatic mammalian cells, which have followed the pattern of molecular-biological developments in the simplest microorganisms. It is intended to demonstrate that large regions of mammalian cell behavior have been opened up to exploration by means of the new and simple techniques which are available today. Many different examples could have been used to illustrate this situation. In order to minimize effort, such examples have often been taken from work in the laboratory of the author and his co-workers, although at least as effective experimental illustrations from other laboratories would have served. The present state of mammalian cell biology has received fundamental contributions from scientists in so many different countries as to make this a truly universal field of human endeavor, most appropriate to studies so close to man.

In the arrangement adopted in this book, the mammalian cell is viewed as a microorganism, i.e., those properties are described which have lent themselves particularly effectively to study by the genetic-biochemical techniques which proved so successful with bacteria. In addition, the last part of each chapter attempts to evaluate some of the human implications of the scientific developments considered. This treatment arises from the conviction that science no longer can afford the luxury of regarding itself as a purely intellectual discipline divorced from human problems. Society today is undergoing a deep-seated revolution which is accompanied by agony for all mankind. Perhaps the most potent factor in these developments is the unbalanced growth and power of science and technology. Study of the content of science must be accompanied by some consideration of its possible actions, and students, as they acquire the tools of science, must at the same time understand the powerful ways in which these tools can affect human life. It is hoped that readers of this book will expand for themselves the possible human implications of the developments described here and that scientists working in these fields will accept the responsibility for helping to enlighten society about these implications so that new scientific advances can be amalgamated into the social structure, with enrichment rather than degradation of human life.

This book represents an arbitrary selection of topics chosen by the author to illustrate some of the new developments which make study of mammalian cells one of the most exciting fields of modern biology. He alone must bear the responsibility for the presence of errors. However it is a pleasure to acknowledge the valuable criticisms and suggestions, provided by many people, and especially those of Drs. J. Bonner, P. Berg, R. Ham, D. Mazia, H. Schachman, G. Stent, and F. T. Kao. Most particular thanks are due Miss Judy Goddard without whose editorial work, and assistance in all stages of this writing, the task could not have been accomplished.

April 1972 *Theodore T. Puck*

CONTENTS

THE MAMMALIAN CELL AS A MICROORGANISM

Genetic and Biochemical Studies *in Vitro*

CHAPTER 1

Growth of the Mammalian Cell as an Independent Microorganism

Much of the explosive development of molecular biology was made possible by the availability of simple techniques for studying the genetics and the genetic biochemistry of microorganisms, such as *Escherichia coli.* (See Appendix I.) The basic operation, the single-cell plating procedure, involves placing a measured number of single cells in a petri dish containing a complete nutrient medium for cell reproduction. An agent, usually agar, is added, producing a stiff jellylike consistency that keeps each cell and all of its progeny together so that the cells multiply to form discrete, visible colonies. This procedure permits several fundamental operations. First, the number of colonies produced can be counted, allowing an accurate measurement of the fraction of cells of any population capable of reproduction to the point of visible colony formation. Second, the exact molecular environment needed for reproduction of single cells can be accurately determined. Third, this procedure permits the determination of survival curves, that is, the number of cells remaining capable of continued reproduction after application of various doses of a physical, chemical, or biological agent. From the shapes of such curves, important deductions about the mechanism of action of the given agent can be drawn.

But perhaps the most important feature of single-cell plating is that it permits a variety of powerful genetic studies. When single cells multiply in isolation, the colonies they form are clones, that is, they consist of individuals derived from the original cell by asexual reproduction. Hence, these consist of the most genetically uniform populations possible. Therefore, mutant colonies can readily be recognized, isolated, and grown up into a large culture, all of whose members will display the mutant characteristic. These mutations then act as markers which permit tracing of the pathways of biological inheritance and their accompanying biochemistry. The single-cell plating technique permits isolation of many kinds of mutations by plating a large cell population under conditions that will not permit growth of the original form but that will permit growth of particular mutants. Thus, actions of mutagenic agents can be studied, and large numbers of mutant cells can be obtained and compared with the original cell in order to illuminate the processes controlling the expression of various kinds of genetic potentialities.

Classical genetic studies of mammals, in general, and of man, in particular, have relied on the normal mating procedure of mammalian reproduction. These present formidable difficulties because the generation time of man is about 25 years (that of *E. coli* is only 25 minutes), and those matings that would be the most illuminating from a genetic point of view cannot readily be carried out in man.

Since a mammal such as man consists of an aggregate of approximately 10^{13} somatic cells, and since presumably each cell contains a complete genome, focusing on the reproduction of individual mammalian cells offered hope that the methods so successful in microbial genetics might also be applicable to populations of mammalian cells. Thus, many kinds of genetic and biochemical studies of mammalian systems would be possible if (1) stable growth of mammalian cells could be instituted in tissue culture without change in genetic constitution, (2) single cells from such populations could be induced to grow into colonies by simple, rapid, and quantitative means, and (3) mutant clones could be recognized, selected, and grown up to form new cell populations. Moreover, the ability to grow clonal cultures should make possible separation of the effects on cellular behavior of genetic constitution, developmental state, molecular environment, and interaction with other cells.

1-1 EARLY STUDIES ON GROWTH OF MAMMALIAN CELLS *IN VITRO*

At an early date, investigators attempted to grow mammalian cells in test tubes where the environment could be more readily controlled and manipulated than in the body. Long-term growth of cells outside the body was claimed to have been accomplished early in the century, but it is now recognized that these experiments are highly suspect, since the cells were nourished by a so-called "embryo extract," which had been prepared by methods known to leave large numbers of whole and viable cells within the final preparation. While this early work may have been erroneous, it stimulated many scientists to turn their attention to the possibilities of tissue culture as a scientific tool for study of mammalian cells.

One of the great difficulties encountered initially was the establishment of cell cultures that would continue to grow for long periods of time. Samples of cells taken from normal tissues of humans and other animals often displayed multiplication for periods varying from several days to several months, after which growth stopped. However, a few laboratories developed vigorously growing cell lines that continued to multiply for indefinite periods in artificial media supplemented by various biological fluids. Among the best known of these were the L cell, isolated by Earle and his associates from a mouse fibroblast, and the HeLa cell, isolated by George Gey from a biopsy on a human carcinoma of the uterus. These cultures, distributed to many laboratories, were very important in helping launch this field on a quantitative basis.

The development of methods for regularly growing massive cultures of some selected mammalian cells *in vitro* opened many new areas of investigation. One of the most important was the possibility of inoculating a dish with a measured number of single cells so that each could reproduce in isolation to form a clone, in a manner analogous to the fundamental operation of quantitative microbiology. Initial experiments of this kind, however, were uniformly negative.

In these experiments, certain mammalian cells were shown to grow with great regularity in mass inocula, but consistently failed to exhibit single-cell growth. Thus it became widely accepted that the single mammalian cell could not grow independently as a microorganism. It was found that a population of at least 100,000 cells was required, in a standard-sized vessel, for growth to occur under the conditions employed. It was inferred that the medium then used to support the growth of cells was incomplete, but that sufficiently large numbers of cells could transform the medium into one adequate for growth of the component cells. This hypothesis was supported by an elegant experiment in which long-term growth was achieved from a

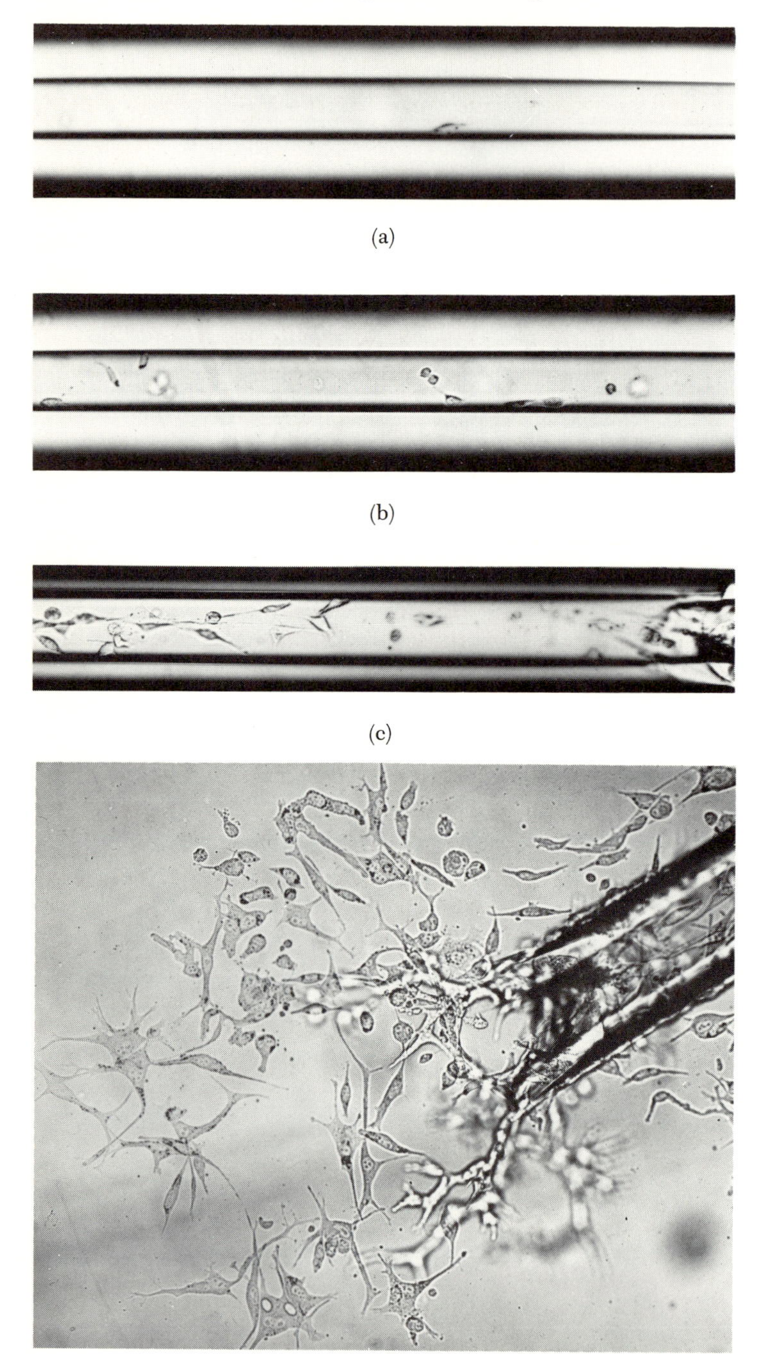

(a)

(b)

(c)

(d)

single cell, provided that it was sealed into a very small capillary (Figure 1-1). The restricted volume of medium enclosed with the cell made it possible for the single cell to "condition" the medium, that is, to transform it to nutritional adequacy. These experiments demonstrated that single mammalian cells could grow *in vitro* into large clonal populations, provided that proper conditions were maintained. The problem remained to devise a simple method of growing hundreds or thousands of single cells into colonies in a single, quantitatively reproducible operation.

1-2 QUANTITATIVE GROWTH OF SINGLE MAMMALIAN CELLS *IN VITRO*

The first successful arrangement for solving this problem consisted of inoculating single cells on top of a "feeder layer" of cells which had suffered irreversible damage to their reproductive function but not to their metabolic activities. Thus, the feeder layer, capable of transforming the medium into a nutritionally adequate one but not capable of reproducing, would not obscure multiplication of the single viable cells added thereafter.

Such a feeder layer was prepared by exposing 10^5 cells to a large dose of x rays in a petri dish (see Chapter 5). An inoculum of 100 untreated single cells was then also added to the dish. The normal cells eventually settled to the bottom of the dish, attached to the glass, and formed colonies. The appearance of the plates after seven days of incubation is shown in Figure 1-2. Within the limits of expected statistical uncertainty, every one of the single cells added to the dish containing the feeder layer produced a large discrete colony of cells. Control dishes containing the feeder layer alone showed no growth of colonies.

Figure 1-1 Early demonstration that large-scale growth could occur in vitro from a single cell provided that the volume of medium was restricted to the point where a single cell could condition it. In this case, a single cell was planted in a small capillary tube and observed after varying periods of time. (a) Original single cell, (b) after 135 hours, (c) after 16 days, and (d) after 28 days. The restricted volume of the capillary has enabled the cell to convert the medium to nutritional adequacy so that extensive growth is obtained. (Photographs courtesy of K. Sanford.)

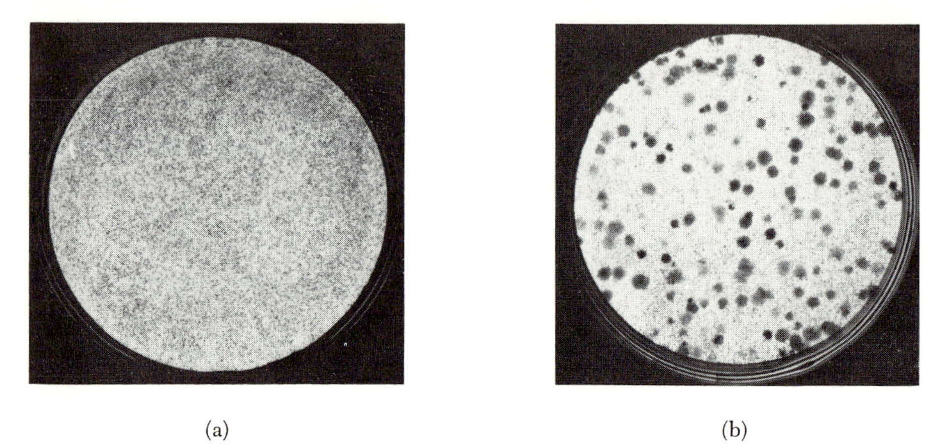

(a) (b)

Figure 1-2 (a) Plate containing 10^5 x-irradiated feeder cells alone. The cells have lost their capacity to reproduce, but still are able to carry out other metabolic functions, including conditioning of the medium. (b) A plate identical to that in (a), but to which have also been added 100 untreated cells. The feeder cells alter the medium, furnishing the necessary molecular environment for clonal growth by the normal cells.

These experiments with the feeder layer made clear that single-cell growth is indeed possible in the mammalian case, provided that a suitable environment is furnished. Therefore, the next step was to develop a medium sufficiently rich nutritionally to promote growth of single cells without a feeder layer. As a starting point, media were used that had been shown by earlier workers to be effective for mass cultures. By empirical experiment, a suitable medium was finally developed utilizing a synthetic mixture of vitamins, sugars, amino acids, and other small molecules, supplemented with a mammalian serum. This medium supported growth of colonies from single HeLa cells without use of any feeder layer (Figure 1-3(a)) and permitted quantitative determination of the plating efficiency of the cell, that is, the fraction of the cell population that forms visible colonies of 50 or more cells. Means were then devised for isolating such individual colonies so that clonal cell cultures could be established. The simplest method consisted of placing over the colony a stainless-steel cylinder whose rim had first been dipped into sterile silicone grease. The grease cemented the cylinder to the bottom of the petri dish so that liquid inside the cylinder could be drawn off by means of a syringe and then replaced by a dilute trypsin solution. The trypsin dissolved the binding of the cells to the glass surface so that the resulting suspension could then be removed with a sterile Pasteur pipette and added to a new vessel. Figure 1-3(b) illustrates the much more uniform colonies which develop when single cells from a clonal culture, established as described, are plated in a petri dish, as opposed to cells from the parental culture shown in Figure 1-3(a). A variety of clones was isolated. Some of these displayed stable genetic differences in their nutritional requirements. (See Figure 1-4.)

With the availability of these techniques, it became possible to cultivate clonal cultures of the HeLa cell indefinitely. Cells were grown as monolayers attached to glass or plastic surfaces (Figure 1-5). When such cultures became confluent, they were harvested as single-cell suspensions by treatment with trypsin, which dissolves the protein cementing the cells to each other and to the solid surface. Aliquots of the resulting single-cell suspension could then be plated in petri dishes for single-cell growth or could be used to seed another set of glass vessels for stock-culture growth. These procedures are identical in principle and only slightly different in practice from the routine procedures of quantitative microbiology.

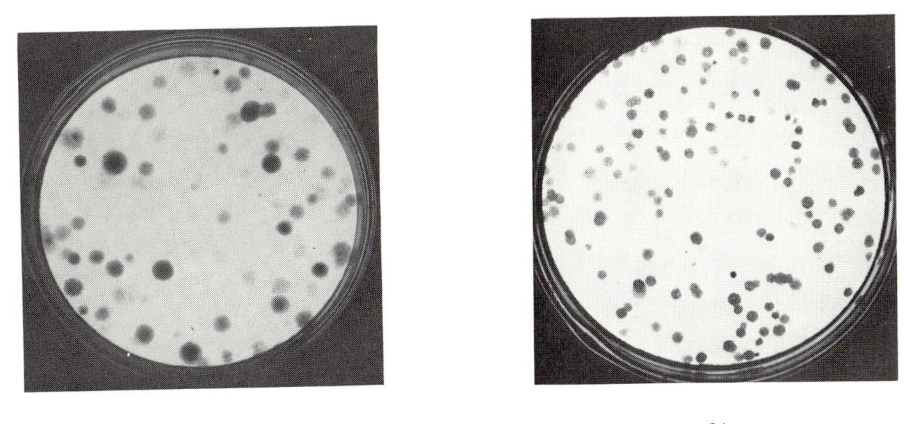

(a) (b)

Figure 1-3 *(a) Colonies that have developed from single wild-type HeLa cells in an improved medium for which no feeder layer is required. The colonies tend to be fairly heterogeneous in appearance. (b) Colonies that have developed from a clone of HeLa cells under conditions identical to those in (a). The clonal nature of the plated cells is responsible for the more uniform appearance of the colonies.*

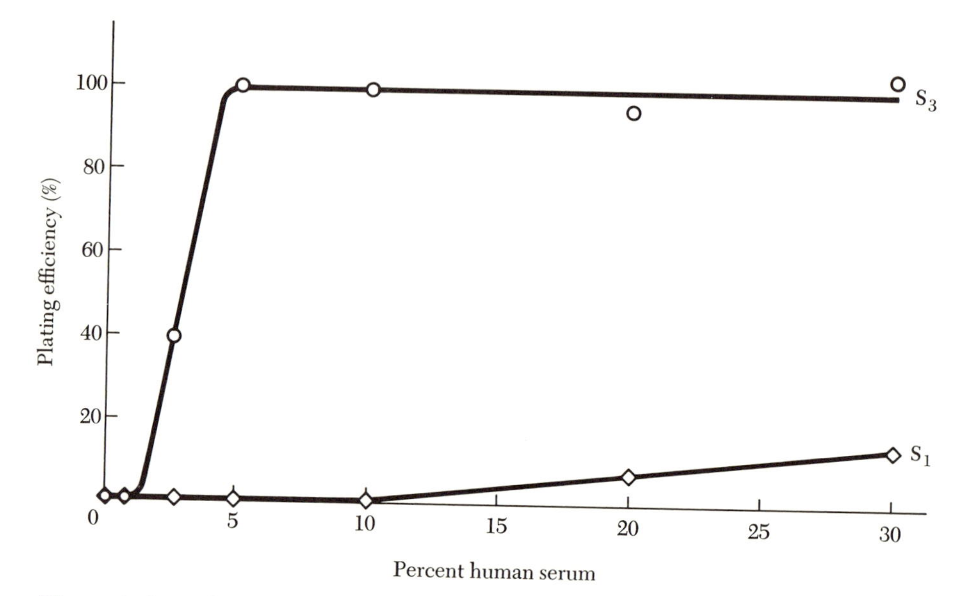

Figure 1-4 *Growth curves for two different clones of HeLa cells demonstrating the different levels of human serum needed for single-cell growth. Other components of the media were identical in both cases.*

1-3 GROWTH OF CELLS FROM ANY HUMAN OR ANIMAL SUBJECT

The experimental developments just discussed were first carried out with the HeLa cell, which is a hardy and efficiently reproducing cell. However, if these techniques were to be widely applicable in studies of mammalian genetics, it would be necessary routinely to initiate new cell cultures from any human subject or experimental animal, to cultivate mass cultures for long periods, and to grow colonies from single cells. In this early period, cell growth from fresh biopsies of mammalian cells was initiated with great difficulty and usually lasted for very short times. Some long-term cultures were established and became widely used for studies of mammalian cells *in vitro;* however, in every case, the chromosomal constitution of such cells was found to be grossly distorted both in number and in structure from that of the original animal.

Attention was therefore devoted to development of manipulative procedures that would involve minimum trauma when mammalian cell suspensions are prepared from tissue biopsies and to the determination of the most efficient nutrient medium and physical conditions of growth. It is difficult for a person familiar only with the relatively rugged growth responses of bacteria and other microorganisms to appreciate the enormous sensitivity of mammalian cells when removed from their natural habitat, which is the result of a long evolutionary development. This sensitivity is illustrated in Figure 1-6, which demonstrates the effect of growing mammalian cells in petri dishes in an incubator supplied with a CO_2 atmosphere that controls the pH of the medium, but with no other special precautions beyond those customary in the growth of microbial organisms.

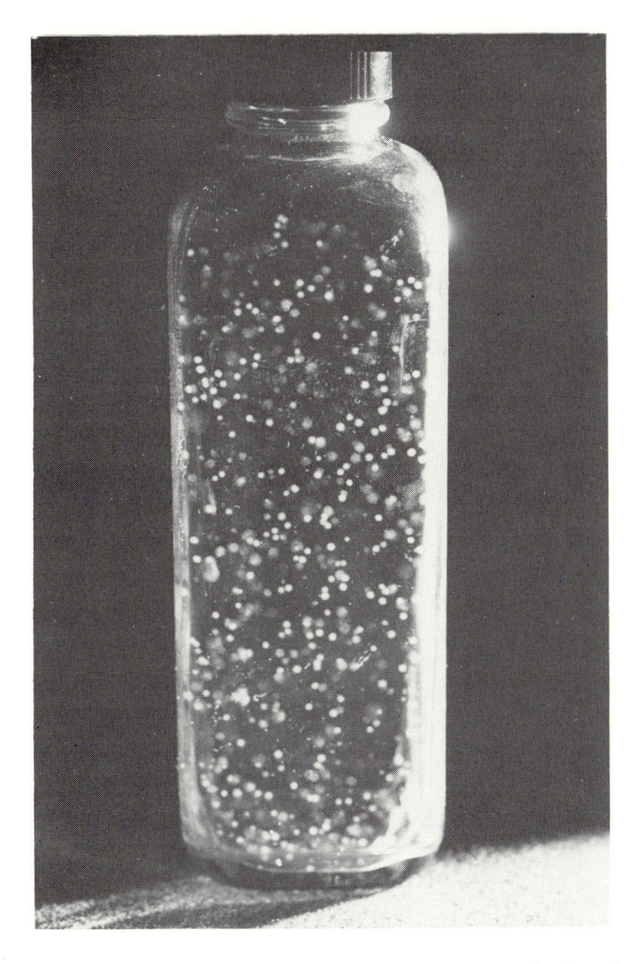

Figure 1-5 *Colonies from single HeLa cells developing on the walls of a glass incubation bottle. Growth can also be obtained in suspension by the use of vessels with walls that resist cell attachment or by rapid stirring.*

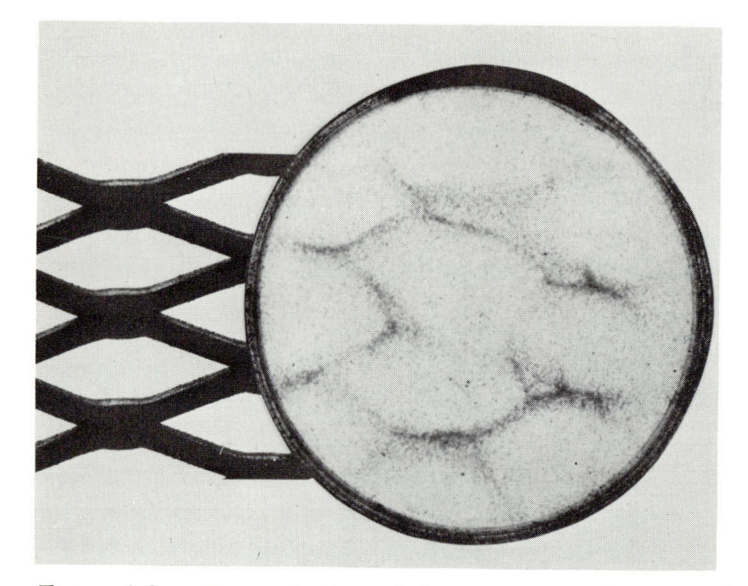

Figure 1-6 *Demonstration of the enormous sensitivity of mammalian cell growth in vitro to small fluctuations in temperature. A petri dish was seeded with an almost confluent amount of HeLa cells, then placed on a metal-grid shelf in a 37° incubator. The incubator doors were opened several times a day during the subsequent incubation, so that the air temperature fluctuated somewhat. It was found that cell growth was heaviest along the lines of the metal part of the grid, which tended to maintain a more constant temperature. The temperature differential between the areas lying directly over the metal and those over an adjoining air space cannot have been very great, but sufficed to produce this marked difference in cell growth.*

With these considerations in mind, methods of cell sampling from various tissues were developed that exposed the cells to the minimum amount of trauma from treatment with chemical agents such as trypsin and from physical manipulations. Electronically controlled incubators were designed to maintain constancy of temperature, CO_2 concentration, and relative humidity which, by controlling the degree of evaporation or condensation of water, determines the osmolarity of the medium. Attention was then turned to the constituents of the medium itself, which were adjusted to maximize the plating efficiency of the single cells obtained directly from mammalian tissues. It was subsequently found that conditions that maximize the growth of single cells from normal tissues are also maximally suitable for their long-term cultivation. Thus, under these conditions, cells from a variety of normal human and animal tissues can be routinely plated as single cells in petri dishes, will grow in long-term culture with a stable chromosomal complement, and will produce colony formation with efficiencies of 30 to 80%. Virtually all of these cultures produce cells of the spindle shape resembling the fibroblasts that are seen *in vivo*, and the colonies developing from single cells of such cultures present an appearance in which bundles of such cells are lined up parallel to one another. While such cultures can be maintained in active growth for periods as long as a year and for approximately 100 generations (during which the cell number could increase by a factor of 2^{100}), and can be stored for long periods in the viable state in the deep freeze, eventually the growth of these cultures may slow down or cease altogether. The reason for this growth cessation has not yet been clarified. Hayflick has suggested that the mammalian cell may be capable of only a limited number of reproductive acts and that this phenomenon is the basis of the aging process.

With the routine establishment of cultures from any individual, definitive studies were possible not only on normal human chromosomes but also on those obtained from persons with particular types of chromosomal and single-gene defects. The rapidity with which these developments found application to problems of human genetic disease surpassed even the most optimistic expectations. These events are discussed in Chapter 2.

1-4 DEFINED MEDIA FOR SINGLE-CELL GROWTH

Use of media containing complex natural substances such as whole mammalian serum is unsatisfactory for analytical studies on cell growth, genetics, and biochemistry. Early studies designed to elucidate the specific nutritional requirements for cell reproduction *in vitro* utilized growth of mass cultures. Under these conditions it was found that most mammalian cells, regardless of their source, exhibited closely similar nutritional requirements. However, when single-cell plating was employed, individual differences between cell types became apparent. Presumably the mass-culture situation permits cooperative effects so that marginal biochemical capacities that are inadequate for growth of isolated cells become capable of supporting reproduction in a large population.

In addition it was found that HeLa and similar cells, which have supernumerary chromosomes, are less demanding nutritionally than cells with a normal or hypodiploid chromosomal number. Single-cell growth of both types, however, requires one or more protein components. Some cells require albumin; all cells so far tested require an interesting serum glycoprotein. This requirement can be satisfied by glycoproteins from a variety of mammalian sera, but the best source is the alpha-globulin fraction of fetal calf serum which consists principally of a glycoprotein called fetuin. Table 1-1 presents the components of a medium that will provide virtually 100% plating efficiency and high growth rate for a representative cell, isolated from Chinese hamster ovary, a cell type that has proved useful in several different kinds of genetic-biochemical experimentation.

Table 1-1 Composition of F12D, the Minimal Medium for the Chinese Hamster Ovary Cell (F12, the enriched medium discussed in Chapter 3, can be obtained from F12D by the addition of nine metabolites)

Amino Acids (16)	*Vitamins and Growth Factors* (9)	*Inorganic Salts* (9)
Arginine	Biotin	KCl
Histidine	Calcium pantothenate	$Na_2HPO_4 \cdot 7H_2O$
Isoleucine	Niacinamide	$FeSO_4 \cdot 7H_2O$
Leucine	Pyridoxine	$MgCl_2 \cdot 6H_2O$
Lysine	Thiamine	$CaCl_2 \cdot 2H_2O$
Methionine	Folic acid	$CuSO_4 \cdot 5H_2O$
Phenylalanine	Riboflavin	$ZnSO_4 \cdot 7H_2O$
Serine	Choline	NaCl
Threonine	1,4-Diaminobutane	$NaHCO_3$
Tryptophan		
Tyrosine	*Carbohydrates and Derivatives* (2)	*Lipids* (1)
Valine		
Glutamine	Glucose	Linoleic acid
Cysteine	Sodium pyruvate	
Asparagine		*Protein* (1)
Proline		Fetuin

The metabolic role of glycoproteins such as fetuin in promoting growth of mammalian cells *in vitro* is still unsolved. The active alpha-globulin fraction of fetal calf serum has been resolved by ammonium sulfate precipitation and column chromatography into several active glycoprotein components. One particularly pure fraction has a molecular weight of 45,000, about 20% of which consists of sugar residues.

Addition of fetuin markedly stimulates growth of all cells tested, and fetuin can be titrated by this action (Figure 1-7). Antisera against fetuin prepared in the rabbit will inhibit the growth of cells *in vitro;* this inhibition is specifically reversed by fetuin. When cells are placed in the defined medium without fetuin, a marked fall in DNA, RNA, and protein synthesis occurs. However, the fundamental mode of action of this protein is still unknown. One of the difficulties in determining the mode of action of fetuin and related substances arises from the instability of the growth-promoting activity. Relatively gentle treatments, such as binding of fetuin to and elution from a chromatographic column such as diethylaminoethyl cellulose, can cause complete and irreversible loss of biological activity, unless certain critical conditions such as salt concentration, pH, and temperature are maintained. These phenomena suggest that biological activity may depend on the presence in the glycoprotein of a detachable small molecular moiety, but so far efforts to establish the existence of such a substance have been unsuccessful. Cells grown *in vitro* have been shown to synthesize small amounts of a glycoprotein that immunologically, at least, is similar to fetuin; this finding may explain why

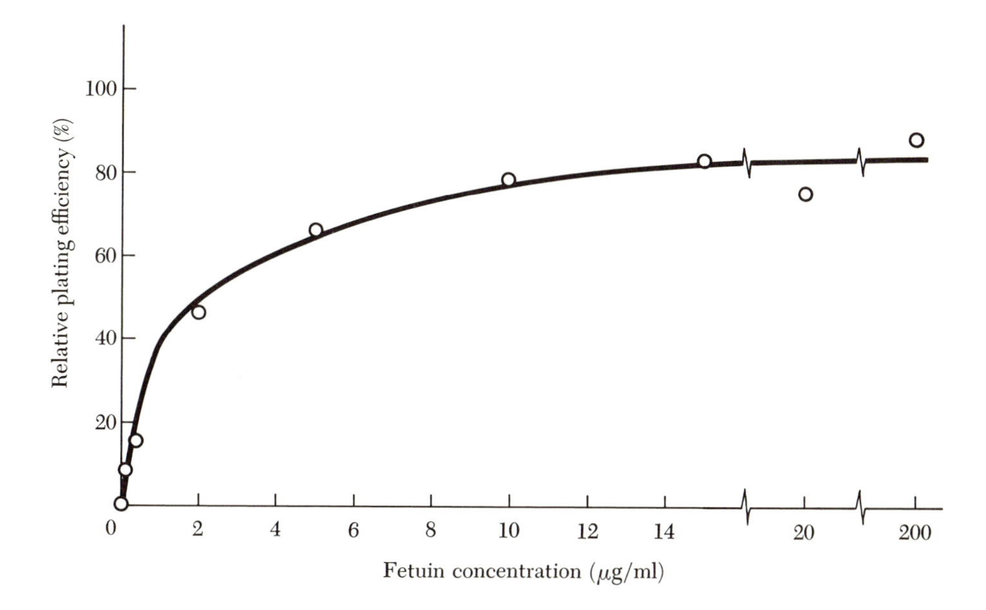

Figure 1-7 Titration curve showing the action of fetuin in promoting single-cell growth of cells from a bovine ovary culture.

some cells can grow at least slowly in a completely protein-free medium.

A defined medium has not been achieved for all mammalian cells. For example, some human cells will not produce colonies in the medium described in Table 1-1, but will do so in media supplemented with a combination of whole sera from the fetal calf and human cord blood. Determination of the exact molecular requirements is a painstaking but rewarding process, as will be seen in Chapter 3 where the production of nutritional mutants is described.

1-5 OTHER ASPECTS OF CELL GROWTH *IN VITRO*

Mammalian cells can be cultivated in two principal fashions. If grown in vessels made of glass or certain kinds of plastic, the cells attach and adhere tightly to the surface, growing as a sheet, which, under certain circumstances, may present patches of three-dimensional growth (Figure 1-5). If prevented from such surface attachment by a high rate of stirring or by the use of vessels whose surfaces resist cell attachment, growth of some but not all kinds of cells can also be attained in suspension. By measuring the increase in cell number under either set of conditions, the average time necessary for one cell generation is readily obtainable. Depending upon the nature of the cell, the medium, and the temperature employed, generation times as low as 8 hours or as high as 40 or 50 hours have been observed. Single mammalian cells will grow into colonies at temperatures ranging from 35°-41°C; however, maximum plating efficiency is obtained in the neighborhood of 37°C.

If single cells are plated in a petri dish and then examined at daily intervals, the number of cells in each developing colony can be counted, and the time course of single-cell growth can then be plotted as shown in Figure 1-8. This simple method of constructing a growth curve demonstrates the similarity between the behavior of these cells and of a growing bacterial population. Thus, there is an initial lag period presumably required for the cells to adapt to the nutrient medium. Thereafter, growth follows an exponential course with high fidelity. The reliability of this procedure is shown by the concordance of the results indicated by the black and hollow circles in the figure, which represent, respectively, values obtained from two similar experiments carried out on the same culture in the same medium but separated by a time interval of one year.

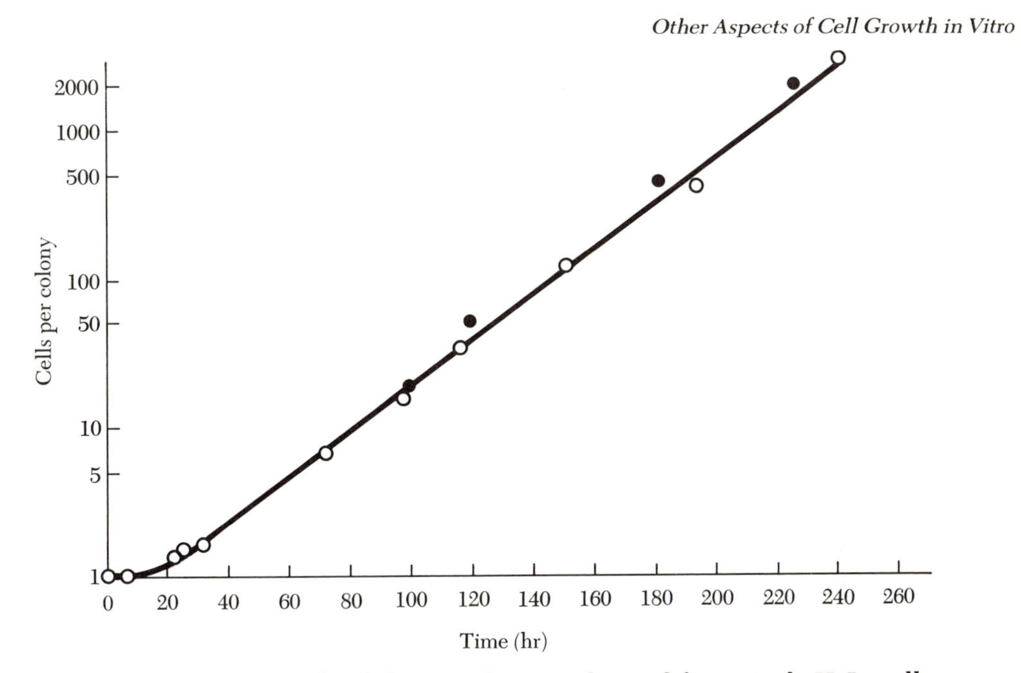

Figure 1-8 ***An example of the growth curve obtained from single HeLa cells grown in a complete nutrient medium. The reproducibility of the technique is evident in that the data from two different experiments carried out one year apart (solid versus hollow circles) are virtually identical.***

The morphological appearance of cells grown *in vitro* is affected by a large variety of factors. One of these is the genetic constitution of the cell—hyperploid cells tend to assume an epitheliallike appearance in the same medium in which normal diploid cells are usually spindle-shaped, that is, fibroblastlike. However, a given cell can assume a variety of different shapes *in vitro* depending on the constitution of the surrounding medium. Fetuin added to cells in an otherwise complete nutritional medium promotes their conversion from a spheroidal to a flattened form with many pointed processes (Figure 1-9) while addition of gamma globulin from an animal immunized against fetuin promotes the reverse change. Varying the concentration of human serum in the nutrient medium can induce cell flattening and formation of less dense colonies in the HeLa cell. Certain carcinogenic viruses and chemical agents convert cells from a fibroblastlike pattern, in which individual spindle-shaped cells line up in parallel fashion, to an arrangement in which the individual cells are less uniform in shape and the overall pattern is much less orderly (Figure 1-10). This conversion appears to be irreversible. However, a mor-

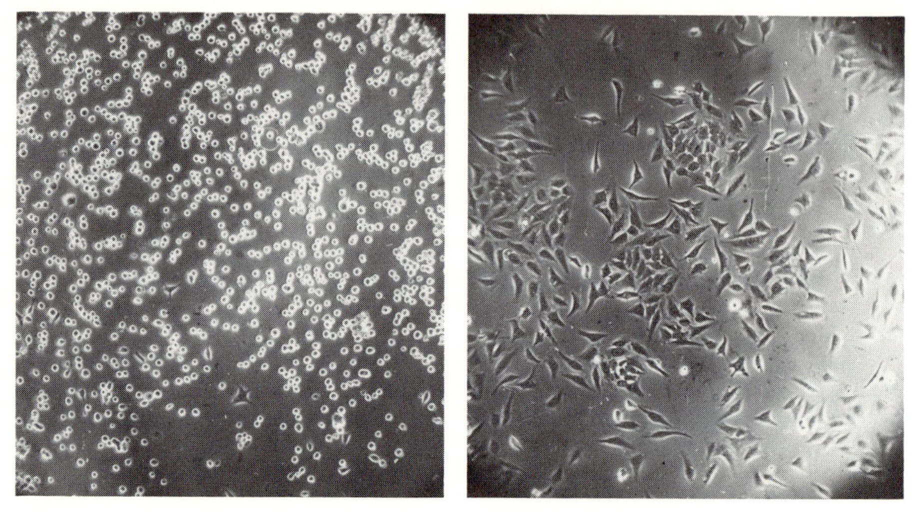

(a) (b)

Figure 1-9 Photomicrographs demonstrating the effect of the growth protein fetuin on the morphology of HeLa cells in vitro (a) in the absence of fetuin, (b) in the presence of fetuin. The stretched cells of (b) can be converted to the rounded cells of (a) by the addition to the medium of the specific antibody antifetuin gamma globulin which neutralizes the action of fetuin.

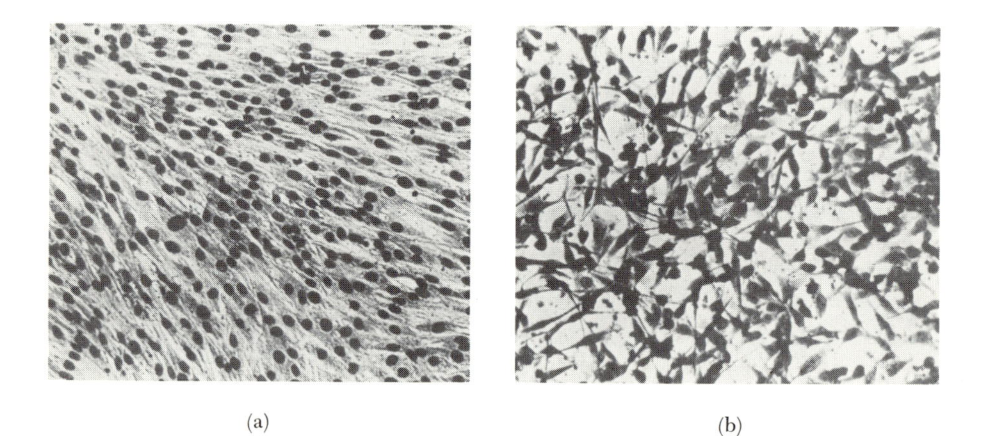

(a) (b)

Figure 1-10 Photomicrographs demonstrating the morphological changes occurring in colonies of Chinese hamster cells treated with the chemical carcinogen benzo(a)pyrene. (a) Normal cells, (b) transformed cells. Tumorogenic viruses produce a similar change. (Photographs courtesy of L. Sachs.)

phological conversion that is similar in certain respects but is reversible can be induced by other chemical agents. One such action involves a metabolic effect of a cyclic adenosine monophosphate (cyclic AMP) derivative. Cyclic AMP is an enormously interesting compound that activates a large number of cellular enzymes and often operates as an intermediary in the production of specific cellular changes by certain hormones. The dibutyryl ester of this compound is often used in experiments because it may be more effectively transported into the cell and is less susceptible to the hydrolytic action of phosphodiesterase than is cyclic AMP.

Chinese hamster ovary cells grown in petri dishes were used in these experiments. Each cell is relatively compact with no particular direction of orientation. The cells will grow not only along the surface of the petri dish but also on top of each other to produce a multilayered colony as shown in Figure 1-11(a). However, if these cells are grown in the presence of dibutyryl cyclic AMP (DBcAMP), they are converted into highly stretched forms which tend to align parallel to their long axes and to associate closely together as shown in Figure 1-11(b). Moreover, under the latter conditions the cells grow only as a monolayer and refuse to produce the kind of multilayered growth exhibited in Figure 1-11(a). Cells which grow only as monolayers are said to be "contact inhibited" since growth stops when the surface is covered with a single layer of cells. It is intriguing that many cells which normally grow in this fashion begin to grow in multilayers after treatment with carcinogenic agents.

A much larger alteration in cell morphology can be produced by an even smaller concentration of DBcAMP if a small amount of the male hormone testosterone is added simultaneously. A clear synergistic effect can be obtained with a combination of 1.5×10^{-5} M testosterone and 3×10^{-5} M DBcAMP even though each agent by itself produces no detectable effect when used at these concentrations.

It has been suggested that cellular morphology is influenced in an important way by microtubules. These are long, linear, rigid structures which are made up of monomeric protein subunits. The organization of the microtubule subunits into larger functional structures is completely inhibited by colcemide, a derivative of the alkaloid colchicine, which specifically binds to the microtubule subunits.

When colcemide, or other agents which specifically inhibit organization of the microtubules, is added to cells simultaneously with

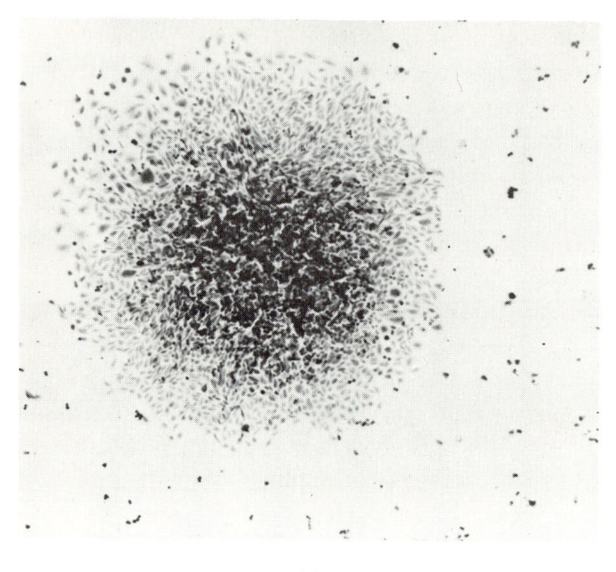

(a)

(b)

Figure 1-11 The effect of a cyclic AMP derivative on the morphology of a colony of Chinese hamster ovary cells. (a) A colony formed from a single cell grown in standard medium. The cells are compact, well separated, and poorly oriented; they have grown in a three-dimensional pattern, thus producing multi-layers of growth. (b) A colony formed from a single cell grown in standard

the elongating agents, the conversion of the epitheliallike form to the contact-inhibited fibroblastlike morphology is prevented. These experiments demonstrate three points: a link between microtubular organization and the action of a cyclic AMP derivative, an interesting synergism between the DBcAMP and a male sex hormone, and the possibility that this morphological transformation, which involves the microtubules, also occurs in the change to the cancerous condition.

Another means for bringing about a reversible conversion to the contact-inhibited state has been developed by Burger and his coworkers. They have demonstrated that cells transformed to the malignant state by a virus could be restored to a normal, contact-inhibited growth pattern by treatment with the glycoprotein concanavalin A. This protein has been shown to cover up specific sites on the cell surface that have become exposed as a result of the viral transformation. The possibilities seem promising that observations such as these may lead to fundamental understanding of the nature of contact inhibition and its role in the malignant process.

The availability of techniques for growing clonal cells *in vitro*, where much greater control over the experimental conditions can be achieved, makes possible precise experiments defining the biochemical processes underlying these and other morphological changes in mammalian cells. Such changes may be of importance in a variety of differentiation processes and in cancer.

1-6 SURVIVAL CURVES

The availability of the single-cell plating technique made it possible to determine survival curves following exposure to a wide variety of physical, chemical, and biological agents. Such curves permit deduction of the number of separate interactions involved in cell killing by viruses and other agents. Two typical curves are shown in Figure 1-12.

The first presents the effect of an animal virus in killing HeLa cells. The linearly logarithmic response to addition of varying amounts of virus to a cell suspension indicates that a single virus

medium supplemented with 10^{-3} M dibutyryl cyclic AMP (DBcAMP). In the presence of this agent, the cells elongate to form the spindle shape characteristic of the mammalian fibroblast. The cells line up in parallel fashion to produce a highly oriented colony and are contact inhibited (i.e., they grow only in a two-dimensional pattern to form a monolayer of growth). Testosterone acts synergistically with DBcAMP in producing this effect.

particle is sufficient to kill each cell. In contrast to this behavior, the response of cells to a physical agent, ultraviolet light, is shown in Figure 1-12(b). Here the survival curve exhibits an initial shoulder. This shape indicates that a single ultraviolet quantum is insufficient to kill any cell, but that a large number of quanta can operate cooperatively to produce cell killing. Various mechanisms may underlie this kind of cell resistance. For example, repair mechanisms may exist, which, until they become saturated, can undo the ultraviolet damage; or multiple cell structures may be present, all of which need to be inactivated before death ensues.

1-7 HUMAN IMPLICATIONS

The life of the mammalian organism begins as a single cell which contains in itself all of the potentialities of the cells that will be derived from it by reproduction; of the tissues and organs that develop from their progeny; of the entire organism that will result from these cells and their various interactions; and, to some degree, even of the societies that will be built through the cooperative efforts of such organisms. Therefore, the study of mammalian potentialities, in general, and of man, in particular, must be carried out at many different levels of organization, which, as has been the case in other regions of science, will supplement each other in powerful ways. This chapter has considered methods of analysis of mammalian cells that involve their reproductive properties, that is, aspects of their behavior that they share with the simplest living cells. In subsequent chapters, we show how studies of this kind, while illuminating the biochemical and genetic events involved in cell reproduction, are also contributing understanding to problems of mammalian embryologic development and to aspects of medical genetics.

The last decade has seen intensive application of cellular and molecular approaches to problems of the mammalian cell. The success that has already been achieved in these areas is unquestionably only a foreshadowing of the great illumination that is yet to come. There is some danger, however, that the success of techniques like these may lead to oversimplified views about the functioning of the mammals and man in particular. Powerful additional techniques will be required to study the multifold problems of mammalian structure and behavior at many different levels, and here, as in no other area of science, it is necessary to attempt synthesis of data from exceedingly

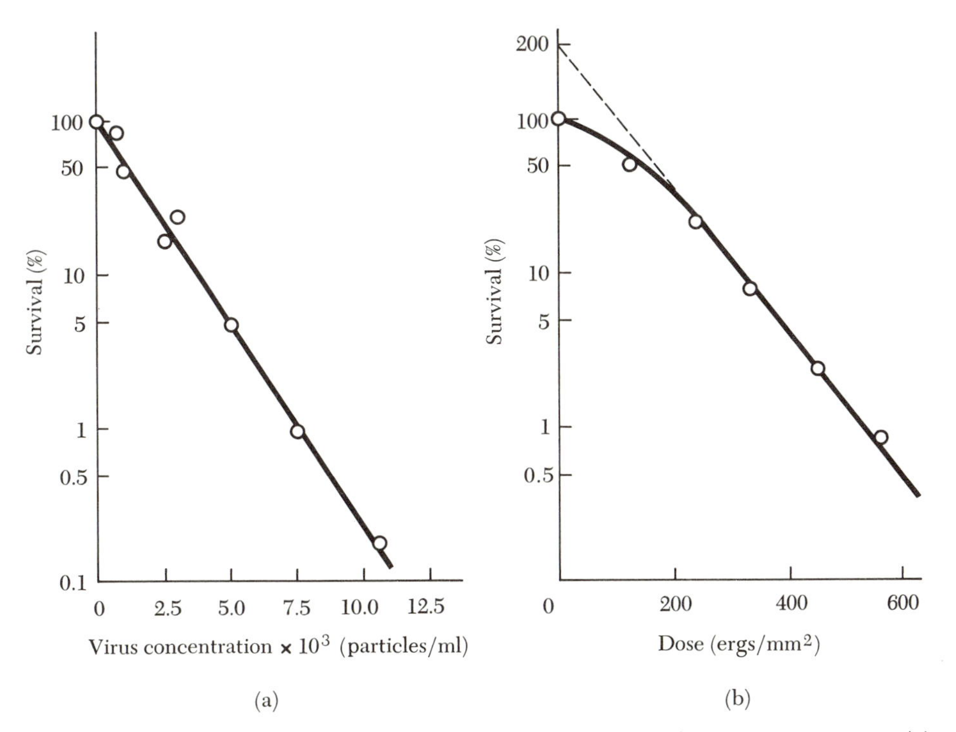

Figure 1-12 The survival of HeLa cells as a function of virus concentration (a) and dose of ultraviolet light (b). In the former case, a one-particle mechanism of killing is demonstrated. Killing by ultraviolet light, on the other hand, exhibits a two-hit behavior. Survival curves of this kind thus offer a simple and reliable means for gaining understanding of the mechanism of action of various kinds of agents on mammalian cells.

diverse kinds of study in order to obtain realistic and useful models for the functions of the mammalian organism as a whole.

1-8 SUMMARY

Techniques have been developed permitting simple, rapid, and quantitative growth of single mammalian cells into large, distinct colonies. Defined media are now available in which every substance required for the continuing reproduction of single cells is identified. These developments set the stage for isolation of mutants and study of the reproductive biochemistry of the mammalian cell in a manner analogous to that which has been so successful in studies of the simplest microorganisms, such as *E. coli.* In addition, growth curves can readily be obtained, the effect of media upon cell morphology and behavior can be studied, and quantitative survival curves can be determined. Study of complex organisms at the single-cell level *in vitro* affords an enormous simplicity not possible in *in vivo* studies. However, synthesis of studies carried out at many different levels of organization is necessary for understanding of the full range of mammalian behavior.

REFERENCES

General

Ham, R. G., and T. T. Puck. Quantitative colonial growth of isolated mammalian cells, *Methods Enzymol.* **5**, 90 (1962).

Puck, T. T. Quantitative studies on mammalian cells *in vitro, Rev. Mod. Phys.* **31**, 433 (1959).

Selected papers

Burger, M. M., and K. D. Noonan. Restoration of normal growth by covering of agglutinin sites on tumour cell surfaces, *Nature* **228**, 512 (1970).

Eagle, Harry. Metabolic studies with normal and malignant human cells in culture, in "The Harvey Lectures, 1958-1959," Academic, New York, 1960.

Hayflick, L. Human cells and aging, *Sci. Amer.* **218**, 32 (1968).

Hsie, A. W., and T. T. Puck. Mammalian cell transformations *in vitro.* I. A morphological transformation of Chinese hamster cells produced by dibutyryl cyclic adenosine monophosphate and testosterone, Proc. Natl. Acad. Sci. **68**, 358 (1971).

Marcus, P. I., and T. T. Puck. Host-cell interaction of animal viruses. I. Titration of cell-killing by viruses, *Virology* **6**, 405 (1958).

Puck, T. T., P. I. Marcus, and S. J. Cieciura. Clonal growth of mammalian cells *in vitro.* Growth characteristics of colonies from single HeLa cells with and without a "feeder" layer, *J. Exptl. Med.* **103**, 273 (1956).

Puck, T. T., C. A. Waldren, and J. H. Tjio. Some data bearing on the long-term growth of mammalian cells *in vitro*, in "Topics in the Biology of Aging," edited by Peter Frohn, Interscience, New York, 1966, p. 101.

Sanford, K. K., W. R. Earle, and G. E. Likely. The growth *in vitro* of single isolated tissue cells, *J. Natl. Cancer Inst.* **9**, 229 (1948).

CHAPTER 2

The Chromosomes of Man and Other Mammals

Study of the genetics of microorganisms involves, for the most part, examination of specific gene mutations and their corresponding effects on the synthesis of specific proteins. Damage or loss involving a large portion of the single DNA chain of *E. coli* causes death. In mammals, the genes are divided among many chromosomes and are present in the diploid condition. Therefore, extensive changes in large parts of individual chromosomes, or even in the total number of chromosomes, can occur without producing lethality. Because each mammalian chromosome may contain thousands of genes, it is obvious that gain or loss of a chromosome or changes in chromosomal structure can seriously affect the structure and behavior of the cell, as well as of the organism that develops from it. Therefore, mammalian genetics must be studied at two different levels: the chromosome and the individual gene.

In contrast to the single DNA chain of *E. coli*, the mammalian cell contains a large number of complex structures, the chromosomes, which exist in pairs so that there are two copies of each gene. Thus, the 46 chromosomes of man consist of 22 pairs of autosomes and 1 pair of sex chromosomes. This doubled condition of the chromosomes confers a biological advantage on the organism in providing a factor of safety against mutation of vital genes. The amount of DNA in each human cell would, if stretched out as a linear double helix, extend to a length of approximately 6 feet, and the total information content in the ordering of the base pairs is incredibly large. In interphase cells the active sequences of DNA engaged in RNA synthesis or self-replication must be maximally extended if the individual base pairs are to be exposed to the surrounding solution. The chromosomes under these conditions cannot be individually identified; they present a diffuse microscopic morphology. At mitosis, however, the chromosomes are enormously condensed so that their length decreases about 10,000-fold, as compared with the interphase condition, and a corresponding increase in cross-sectional area occurs. The compacted structures so produced can be precisely distributed among the two daughter cells. Each mitotic chromosome assumes a definite and readily described shape, which can be used for its cytological identification, as is discussed in Section 2-2.

2-1 HISTORY

The chromosomes of mammals, in general, and of man, in particular, have been of extraordinary interest ever since they were first observed. However, understanding of their most elementary properties was delayed for decades because of lack of suitable techniques. Through the almost simultaneous development of several new laboratory approaches, an explosive series of discoveries was initiated, which has made significant contributions to the fundamental genetics of mammals and other multicellular organisms and has had dramatic application in medicine.

In the early decades of this century, methods of chromosomal determination of mitotic and meiotic cells of a variety of plants were developed, and plant cytogenetics became a rapidly expanding scientific field. The human chromosomes, however, are smaller and more numerous than those of many plants, and great difficulty was experienced in separating the human mitotic chromosomes from each other. In addition, it was difficult to obtain cells from the mammalian body in which mitotic figures were sufficiently numerous to permit chromosomal analysis. In the case of man, abortive fetuses and testicular biopsies were employed for such studies, but these tissues were difficult to obtain. Finally, the delicate structure of mammalian cells made it extremely difficult to obtain complete sets of undistorted mitotic chromosomes. Nevertheless, with the crude and laborious techniques available, careful enumeration of the human chromosomes was attempted. In 1926 it was concluded that the most probable value of the human chromosomal number was 48, a number apparently confirmed by subsequent studies in several laboratories. Confidence in this number developed to the point where it was widely quoted in textbooks and for 30 years was generally accepted as the human chromosome number.

The breakthrough in study of the mammalian chromosomes resulted mainly from two lines of development. First, the introduction of methods for growing cells in tissue culture made possible examination of chromosomes from any person or experimental animal and control of the cellular environment, so that effects of agents and changing conditions could be investigated. Second, simplified techniques, such as the use of hypotonic saline to cause gentle swelling of the cells, made possible separation and characterization of the mitotic chromosomes.

In 1956, a short paper by Tjio and Levan electrified the scientific community and initiated the present era of new discovery in human chromosomal genetics. These investigators utilized short-term cultures of biopsies obtained from human abortus tissue and published pictures, unprecedented in their clarity, of the human chromosomes. The chromosome number was 46. However, skepticism was maintained in many quarters about these results, which differed from those that had been accepted for so many years. Shortly thereafter, a group of English cytogeneticists used improved techniques to examine human testicular biopsies, and again these experiments revealed only 46 chromosomes. A claim was then advanced that the chromosome number of man could equally well be 46, 47, or 48. However, by means of the techniques for growing cells *in vitro* described in the preceding chapter, which had just become available, it was possible to take cultures from the skin of large numbers of human subjects. Such studies, undertaken in several laboratories, demonstrated with overwhelming clarity that the normal chromosome number of man is indeed 46. In a typical study, mitotic figures in over 1800 cells originating in seven different organs of 13 different patients were carefully examined. All except 2 cells had a chromosome number of 46. Pictures of human chromosome preparations obtained from such tissue cultures are shown in Figure 2-1.

In 1962 it was demonstrated that by use of cultures taken from the lymphocytes of the blood, the time necessary for chromosomal determination could be shortened from one week to 48 hours. Unlike the skin-culture technique, this methodology does not permit long-term cultures to be established, because these cells undergo only a few

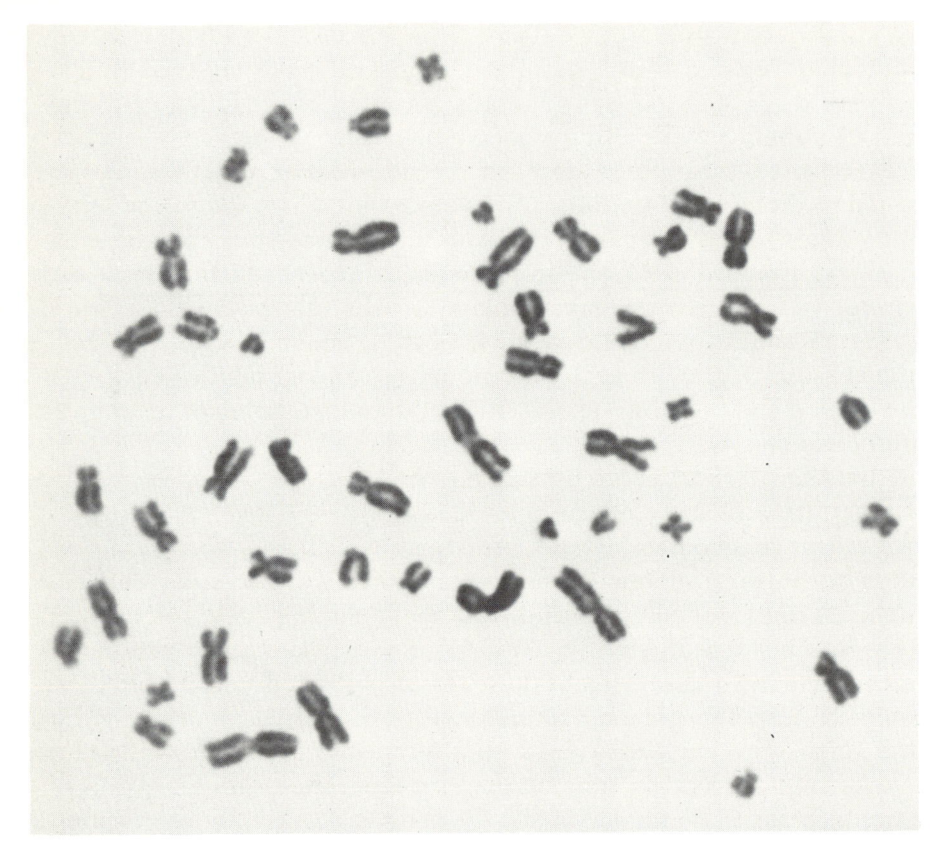

Figure 2-1 *A sample specimen of normal human male chromosomes prepared from skin grown in tissue culture. Modern techniques produce clearly separated and defined chromosomes, which permit counting, measurement, and identification.*

reproductive divisions. However, the increased convenience which it permits has been of tremendous importance in clinical applications. Methods for simplifying the procedure still further are urgently needed because of the necessity for large-scale chromosomal determination in human genetic studies and in medicine. There is also great need to increase the degree of resolution far beyond the present limits. At present writing, computers can examine photographs of mitotic figures almost, but not quite, as well as skilled technicians. It is to be hoped, however, that within several years automated scanning techniques will decrease the necessary time and increase the resolving power of chromosomal analysis.

2-2 STRUCTURE OF THE MITOTIC CHROMOSOMES

Establishment of the number of human chromosomes was followed by measurement of their length and centromere position and by their classification. The longest metaphase mitotic chromosome in man is 6.8 ± 1.4 μ while the shortest is 1.36 ± 0.31 μ. Some of the chromosomes differ in length from each other by only 10%. When one considers the various artifacts that may affect the size and shape of a chromosome in a mitotic cell, such as stretching, bending, folding, or uneven contraction in the course of fixation, it is easy to see the difficulty involved in accurate identification of each member of the chromosomal complement. An International Conference on the Human Chromosomes was held in Denver in 1960 to determine whether enough agreement existed in the results obtained by various workers to justify proposal of a formal system of classification of the human chromosomes. The scheme adopted at this conference is shown in Figure 2-2. Two subsequent international conferences, held in London in 1963 and in Chicago in 1967, confirmed the validity and the usefulness of the Denver system of classification.

The various chromosome groups are indicated by letters, while the individual chromosome pairs are numbered. Neighboring chromosomes in the 6-12 group are particularly similar to each other in size and shape. Hence this group is the one most difficult to resolve. The relative lengths of the different chromosomes and the ratios of their long to short arms are indicated in Table 2-1. At present, length measurement is the chief operation by which the chromosomes are characterized, although other features are helpful. It is to be hoped that, instead of linear measurements, the DNA content of each arm of each chromosome will eventually be measurable directly so that resolving power can be increased and artifacts eliminated. One of the most interesting features is the fact that the X chromosome, which is one of the larger chromosomes and contains approximately 6.3% of the haploid DNA content, is 3 times the size of the Y chromosome. Recently, fluorescence studies have begun to yield patterns aiding in chromosomal identification. Dyes that bind to DNA by intercalating between base pairs have been found to yield fluorescence bands that have a high degree of chromosome specificity. Thus when male cells are treated with appropriate dyes, the Y chromosome fluoresces strongly even in interphase nuclei. At the present time, chromosome

Idiogram of human male

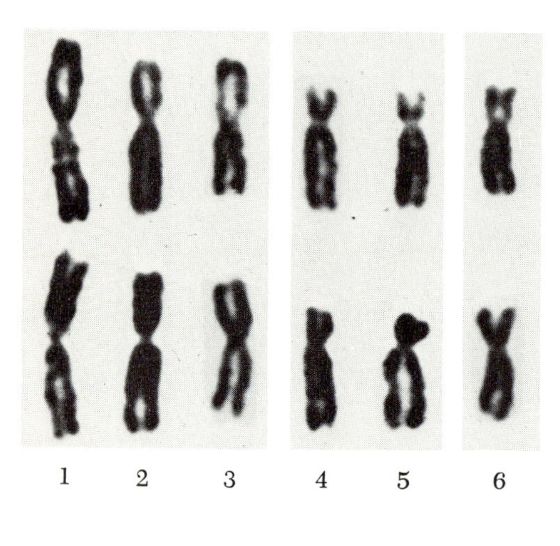

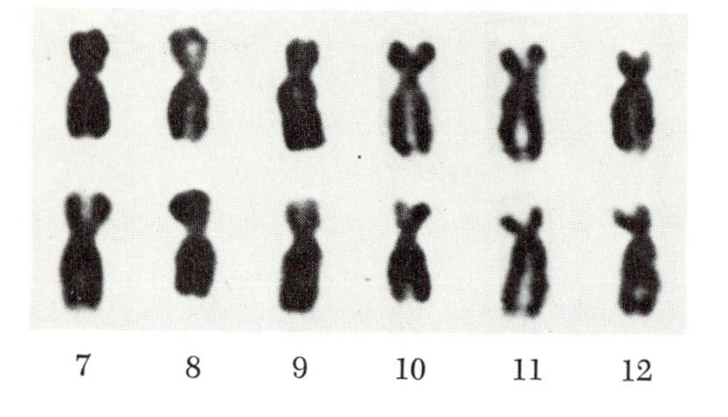

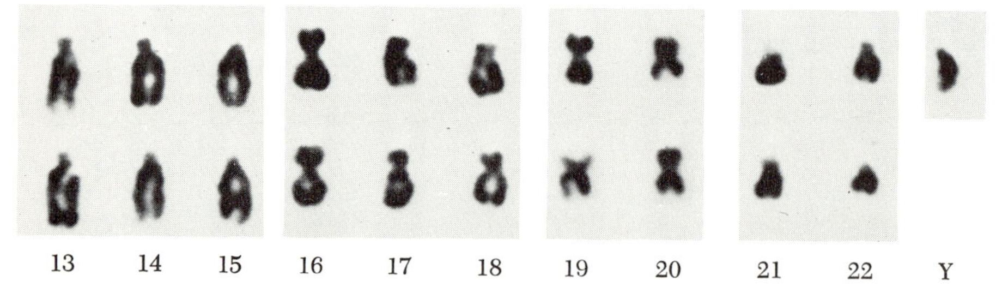

Figure 2-2 Classification of the normal human male chromosomes in accordance with the internationally accepted system. The groups are sometimes given letter designations as follows: (A) 1-3, (B) 4-5, (C) 6-12 plus X, (D) 13-15, (E) 16-18, (F) 19-20, (G) 21-22 plus Y. The female chromosomes are identical, except that the sex chromosomes consist of two X's instead of one X and one Y.

Table 2-1 Classification of the Human Chromosomes According to Their Lengths and Arm Ratios

Group	Chromosome Type	Mean Length Index in Percent*	Ratio Arms Long/Short	
A	1	9.6	1.08	
	2	8.7	1.56	
	3	7.4	1.20	
B	4	6.8	2.85	
	5	6.2	3.18	
C	6	5.8	1.69	
	7	5.0	1.31	
	8	4.7	1.50	
	9	4.7	1.92	
	10	4.6	2.40	
	11	4.6	2.78	
	12	4.5	3.13	
	X	6.3	1.94	
D	13	3.7	8.00	(Satellited)
	14	3.4	7.33	(Satellited)
	15	3.1	10.50	(Satellited)
E	16	3.4	1.78	
	17	3.1	2.83	
	18	2.6	3.75	
F	19	2.3	1.43	
	20	2.2	1.29	
G	21	1.9	3.67	(Satellited)
	22	1.8	3.33	(Satellited)
	Y	2.0	—	(Satellited)

*100 times the chromosome length divided by the sum of the lengths of the haploid autosomes.

pair 22 can be distinguished from pair 21 only by virtue of its weaker fluorescence under appropriate experimental conditions.

The mammalian chromosomes are large, highly complex structures containing even more protein than DNA and about 20% as much RNA as DNA. The protein fraction is rich in histones, which contain

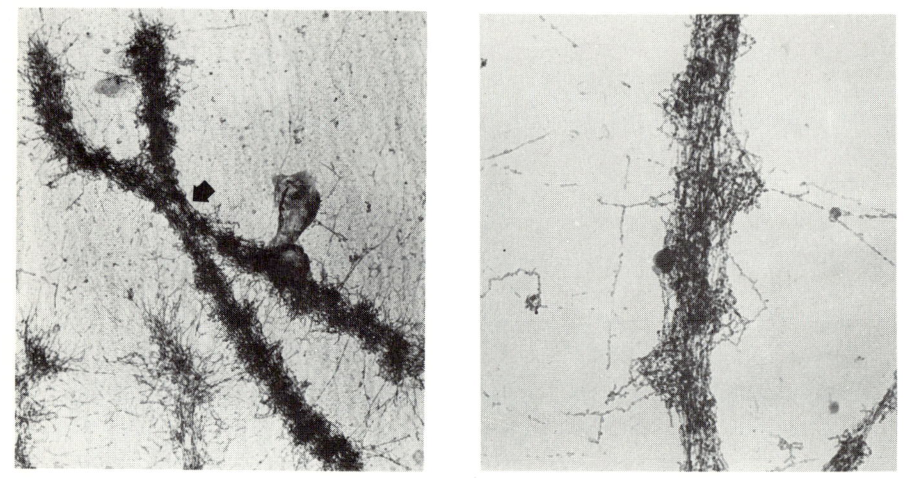

(a) (b)

Figure 2-3 Electron micrographs of human chromosomes illustrating various levels of complexity in their fine structure. (a) Untreated chromosome, consisting of large, closely packed fibers, approximately 240 Å in diameter. Some fibers run longitudinally along the length of the chromosome, and others cross to the sister chromatid either in the centromeric region or further along the chromosome arm. The different patterns of fiber arrangement result in the formation of a "hole" (arrow) at the centromere. The pattern of knobs suggests a spiral structure. (b) Portion of a chromosome arm after treatment with low concentrations of trypsin. Thinning and elongation of the chromosome arm has occurred, but the thickness of the individual fibers is unchanged, suggesting that a cementing material holding the fibers in a compacted condition has been dissolved. A spiral structure is also indicated in this state. (c) Chromosome treated with a more concentrated solution of trypsin. While some fibers of the original size remain, a new, much thinner fiber has also been produced (arrows) of dimensions approaching those of the DNA double helix. (d) Chromosome treated with DNAse shows massive loss of continuity of the chromosome as a whole. (Photographs courtesy of D. Moore.)

relatively large amounts of positively charged amino acids. Recently, electron microscopy has begun to produce new, detailed structural information about the mitotic chromosomes. Electron micrographs of dried and fixed preparations are shown in Figure 2-3. These demonstrate the existence of fibers, approximately 240 Å in diameter, some of which run uninterruptedly along the length of the chromosome, while others may cross to the sister chromatid at the centromere and either continue along the length of the chromosome or reverse

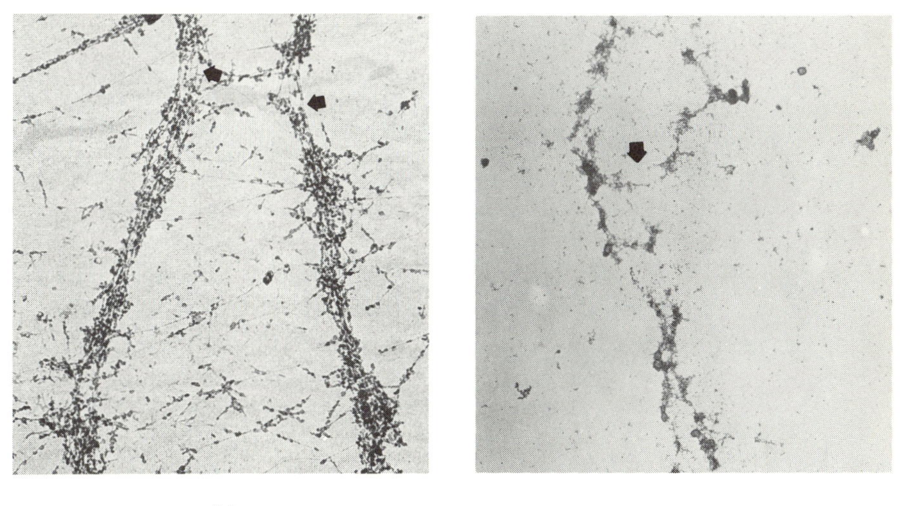

(c) (d)

direction and double back along the homologous sister arm. Still others pass from one chromatid to the other at regions other than the centromere, forming bridges between the sister chromatids. These patterns of fiber arrangement result in the formation of a hole at the centromeric region as seen in Figure 2-3(a). If the chromosomes are treated with low concentrations of trypsin, the fiber structure remains intact. However, the chromosome arms become thinner as though cementing material around the fibers had been dissolved. When partially separated in this way, the fibers reveal a secondary clumped structure that often exhibits a spiral configuration as seen in Figure 2-3(b). After treatment with higher concentrations of trypsin, a new, much thinner fiber appears with dimensions approximating that of a DNA double helix (Figure 2-3(c)), but still maintaining the linear integrity of the chromosome. Digestion with DNAse, however, results in massive disintegration of the chromosome, with destruction of the continuity of the chromosome as a whole as well as that of the individual fibers (Figure 2-3(d)).

The following model is consistent with these observations: The chromosomal DNA is packed into large fibers of 240-Å diameter by means of cementing protein molecules. These fibers are then tightly stacked in parallel fashion by means of binding protein to form the body of the mitotic chromosome; transverse and looped fiber arrangements help maintain the overall structure. The fundamental continuity of the structure depends primarily on the DNA chain. Thus, treatment with proteases first separates the large fibers from each other and then decomposes them, leaving the DNA chain exposed; but complete disintegration occurs only if the DNA chain itself is fragmented as by DNAse.

The condensation of the mitotic chromosomes appears to follow a telescoping arrangement, since an almost continuous progression from prophase to metaphase is observable in which chromosomes progressively become shorter and thicker (see Figure 6-9). The various intermediate states may represent structures containing different numbers and kinds of cementing protein molecules, which make the attachments that fold the DNA bundles into the supercoiled configuration.

2-3 CHROMOSOMES AND HUMAN DISEASE

Early investigators of the human chromosomes of course expected that eventually these studies would have implications for medicine. However, no one was prepared for the magnitude of the impact on the understanding of human genetic disease which the chromosomal delineation produced within an extremely short time. Explanations for some of the classical medical mysteries were suddenly forthcoming, and realization of the serious human health problem caused by chromosomal aberrations was made clear. Such studies have already thrown light on some human developmental processes and promise to be even more illuminating in the future.

Anomalies in both chromosomal number and structure are observed in man. These produce diseases which are most serious when such aberrations are uniform, that is, when they are exhibited by every cell of the body. In this case, the initial error presumably was present in the fertilized egg itself. If the error developed after the first cell division, some of the body cells will show the anomaly and others will not, in which case the individual is called a mosaic. The extent of the resulting pathology varies with the number of cells exhibiting the anomaly. When a sufficiently large number of the cells displays the error, a serious disease may be present in the affected individual, his offspring, or both.

The chromosomal number and structure are constant in the somatic cells of all the body tissues. No differences have been demonstrable in normal human subjects of diverse racial constitution. The first association of chromosomal abnormalities with human disease was made by investigators in Paris in 1959. This epoch-making discovery demonstrated that patients with Down's syndrome (mongolism) possess 47 chromosomes in their somatic cells. The extra chromosome is a chromosome 21, producing a trisomic condition of that chromosome (Figure 2-4). The cause of Down's syndrome had constituted one of the enigmas of medicine for which many hypotheses had been previously proposed. Other conditions associated with specific chromosomal anomalies were described by various investigators

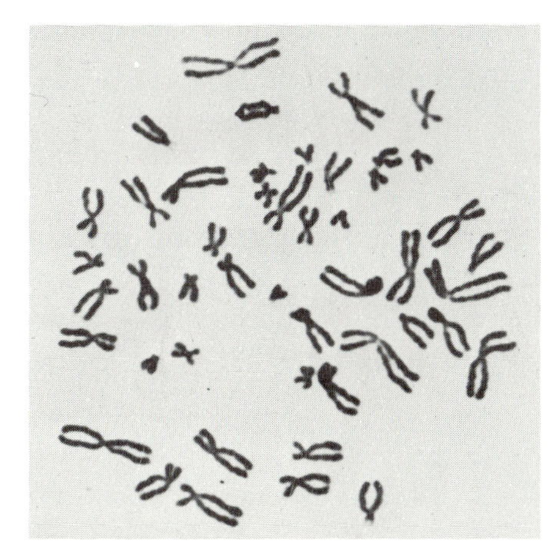

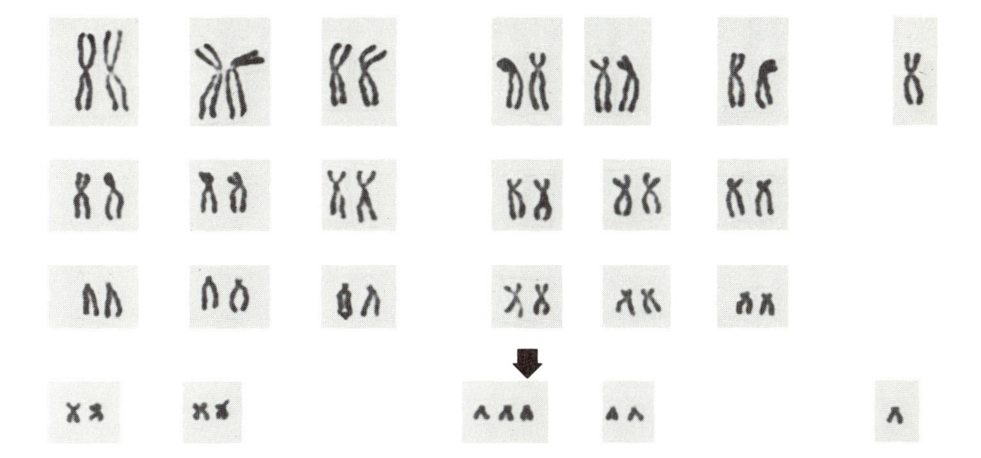

Figure 2-4 Karyotype of chromosomes from a patient with Down's syndrome illustrating the trisomic condition of chromosome 21 (arrow).

within months after this first announcement. At the present time, at least 40 specific disease-producing chromosomal anomalies have been well documented. Moreover, the estimates of the frequency of occurrence in man of these different conditions reveal that approximately 1% of all human live births are accompanied by chromosomal aberration. All of these chromosomal errors are accompanied by deep-seated pathology, and more than two-thirds of the affected persons are mentally deficient, so the health problem involved is enormous. For example, 1% of the male population of institutions for the mentally deficient is accounted for by a single chromosomal anomaly, that of Klinefelter's syndrome. Appendix II summarizes the more common human chromosomal aberrations and their associated pathologic conditions. It is noteworthy that one form of leukemia is definitely linked to a specific chromosomal aberration.

The tragic consequences of these chromosomal anomalies in man point up the need for much greater understanding of the intimate details of human reproduction and development. The most common cause of chromosomal aberration is nondisjunction, that is, a failure

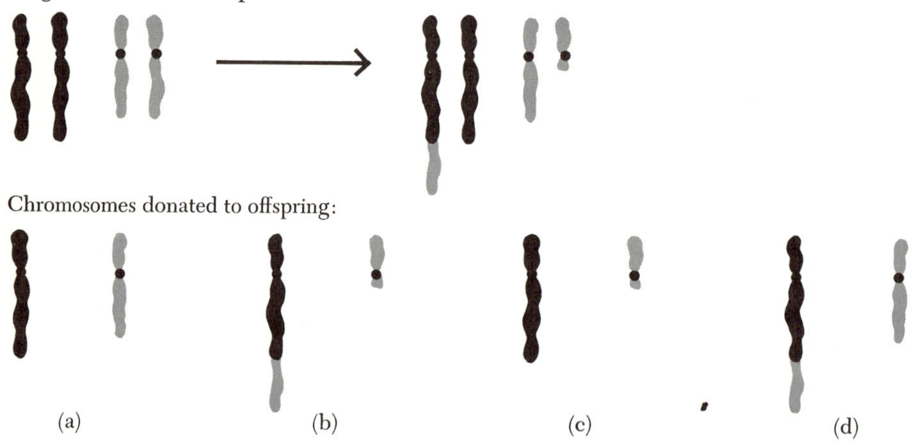

Figure 2-5 Diagram illustrating the four possible ways in which two chromosomes that have been involved in a balanced translocation can be distributed to the offspring. (a) Both of the chromosomes are normal. (b) Both of the chromosomes are abnormal and a balanced translocation results. (c) One chromosome is normal, the other is deficient. Thus a partially monosomic condition results which is often lethal. (d) One chromosome is normal; the other contains an additional fragment which results in a partially trisomic condition. The probability of incurring any one of these conditions is 1/4.

D/D translocation

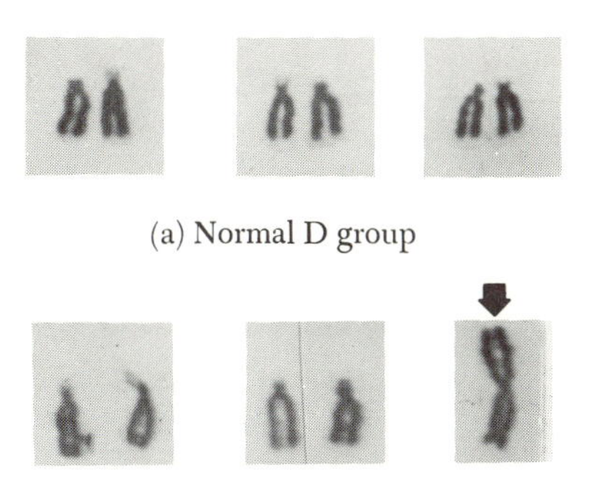

(a) Normal D group

(b) Balanced translocation

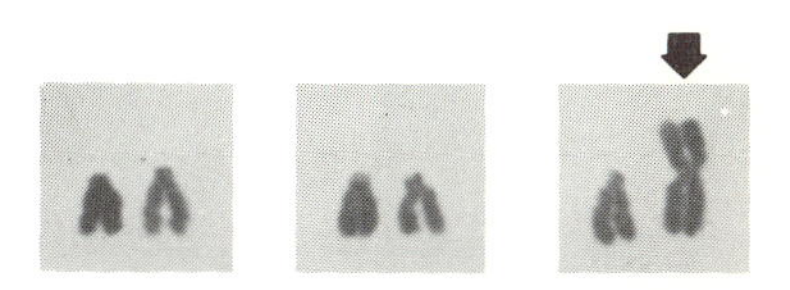

(c) Unbalanced translocation

Figure 2-6 (a) The three pairs of chromosomes comprising the normal 13-15 or D group. (b) Chromosomes of the D group in which a translocation has occurred between two of the chromosomes by means of a centromeric fusion (arrow). The remaining karyotype is normal. No significant amount, if any, of the chromosomal material has been lost or gained, and the subject appears to be completely normal. (c) Chromosomes of the D group obtained from a daughter of the subject in (b). In addition to the normal complement of 23 chromosomes obtained from the mother, the child has received 22 normal plus one aberrant chromosome from the father (arrow). Unlike the father, then, the child suffers growth defects due to the presence of the extra chromosomal material contained in the one aberrant chromosome. Note that in these two individuals, both carrying the same abnormal chromosome, the one with the abnormal number of 45 is clinically normal, while the one with the normal number has suffered severe developmental aberration. Thus an abnormal number of chromosomes in itself is not sufficient to cause developmental defects.

of either the human sperm or egg to receive an exact set of the haploid chromosomes or of a mitotic cell to receive an exact diploid set early in development. Presumably this failure is most often due to an error in the spindle operation. The second general type of aberration is the translocation, which is a consequence of chromosomal breakage and an abnormal rejoining of the fragments. If there has been no loss or gain of chromosomal material, such translocations may be present in an individual without producing any disease symptoms. Such situations are called balanced translocations. However, when such an individual mates with a normal person, the possibility arises for the production of unbalanced translocations in the offspring with resulting abnormal development (see Figures 2-5 and 2-6).

If, after breakage, parts of one or more chromosomes fail to

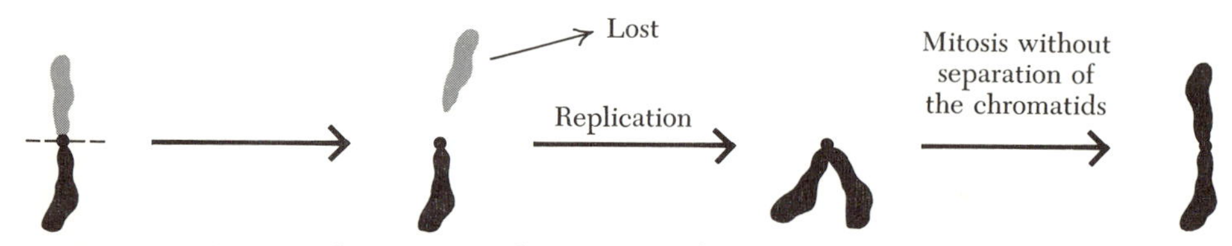

Figure 2-7 Isochromosome formation. Chromosome breakage near the centromere causes loss of one arm. The remaining arm replicates but if the two sister strands stay together throughout mitosis a bi-armed chromosome with two identical arms is produced.

become rejoined to centromere-containing segments, these pieces will be lost to the subsequent progeny of this cell and deletions will result. Finally, if a break occurs exactly at the centromere, a metacentric structure known as an isochromosome may result, in which both arms are genetically identical (Figure 2-7). Occasionally special kinds of other structural abnormalities have been described, such as enlargements of the satellites of chromosomes 13-15 and 21-22 (Figure 2-8).

2-4 SOME IMPLICATIONS OF THESE ANOMALIES

One of the earliest facts to emerge from such studies in man was the recognition that the great majority of chromosomal aberrations in living persons involved at most only seven specific pairs out of the total karyotype. These critical chromosome pairs are 13, 14, 15, 18, 21, 22, and the sex chromosomes. This situation could be explained on the basis of two different hypotheses. One might postulate that these particular chromosome pairs have some structural or functional tendency to participate in processes leading to production of aberrations. Alternatively, it could be assumed that the tendency to participate in such anomalies is more or less randomly distributed among the different chromosomes, but that those chromosomes that rarely or never are found to have participated in nondisjunction usually produce lethality *in utero.*

Evidence discriminating between these two hypotheses is as follows:

(1) In human subjects, translocations have been found that involve virtually every chromosome pair rather than just the critical seven. Therefore, these other chromosomes can participate in basic aberrational processes.

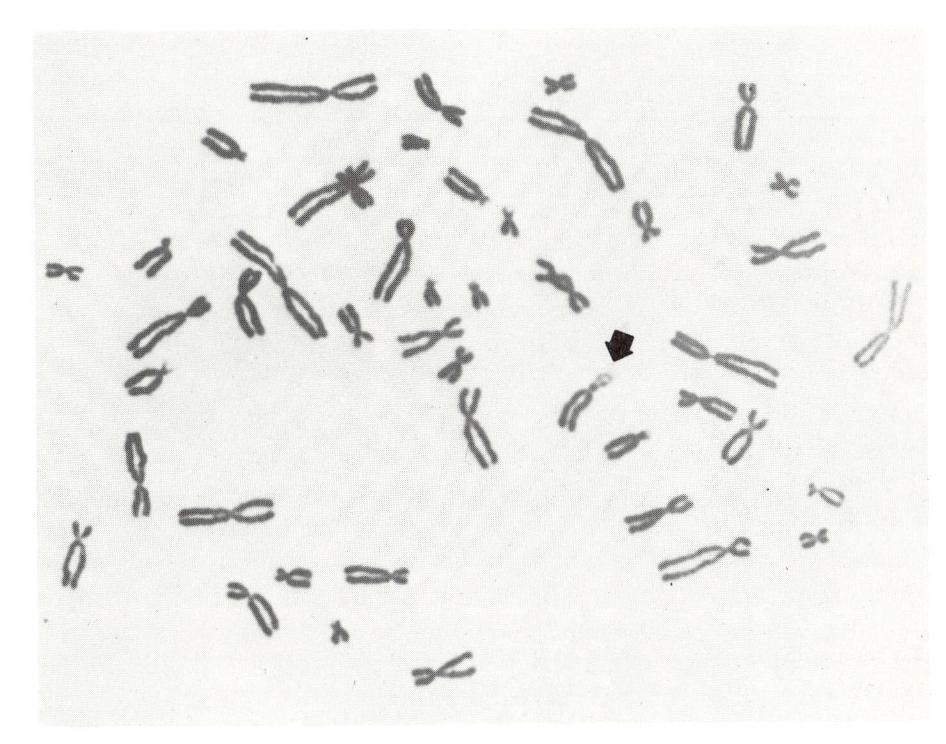

Figure 2-8 *Mitotic chromosomes prepared from cells obtained in the amniotic fluid. The preparations demonstrate the presence in the fetus of an abnormal chromosome, in the 13-15 group, which is characterized by enlarged satellites (arrow).*

(2) There is no evident structural feature that could explain why all the chromosomes should not participate in aberrational processes. Thus the X chromosome, which is one of the largest chromosomes and is almost metacentric, is a frequent participant in these anomalies as is chromosome 21, which is one of the smallest chromosomes and is acrocentric. Chromosomes 6 and X are so similar morphologically that their cytological discrimination is difficult. Yet no case of monosomy or trisomy of chromosome 6 in a living human has been reported.
(3) Anomalies involving large numbers of the chromosomes of the human karyotypes are found in abortive fetuses that have progressed to the stage where chromosome delineation can be performed. Presumably, then, abortions that terminate development at an earlier stage could then contain even more chromosomal anomalies. Further-

more, even for the six autosomal pairs in which aberrations are reasonably frequent in human populations, the affected persons often die at relatively early ages. Thus a high lethality appears to be associated with most known autosomal anomalies.

(4) If one calculates the expected frequency of abortion resulting from chromosomal anomalies, a figure of approximately 10% is obtained (see Appendix III). This figure agrees well with that estimated as the percentage of human conceptions that result in spontaneous abortion through a fetal insufficiency, that is, not connected with physiological malfunction in the mother.

2-5 DEMONSTRATION THAT SPECIAL CHROMOSOMAL REGULATORY PROCESSES EXIST

One of the most surprising developments arising from studies of human chromosomal anomalies was the realization of the great dependence of developmental processes in the mammalian organism on biochemical balance factors. Consider the case of Down's syndrome. In classical genetic theory it has been accepted that one normal gene is usually required for each function. In diploid cells, each gene is present in duplicate, and the presumption has been that this doubling represents a factor of safety introduced by the evolutionary process. Therefore, if a trisomic chromosome were present, certain genes would exist in triplicate, and such an individual might have been expected to be essentially normal with possibly even some advantage. On the contrary, despite the fact that chromosome 21 is one of the smallest chromosomes, containing only about 2% of the total human DNA, the presence of an extra chromosome 21 produces a human baby, in which normal intelligence can never develop and in which certain disturbances of the eyes, the hands, the heart and occasionally the blood-forming organs may be present. The devastating effects of this and other chromosomal trisomies are summarized in Appendix II. The fact that trisomies of most of the human autosomes presumably result in fetal death has already been discussed. Monosomy on the other hand is almost never seen in human autosomes and presumably therefore is almost universally fatal. Deletions of a part of a chromosome are also serious, even when the homologous chromosome is normal. For example, chronic myelogenous leukemia, a malignant disease, is associated with the absence of a small fraction of one member of chromosome 22 (Figure 2-9). The total amount of

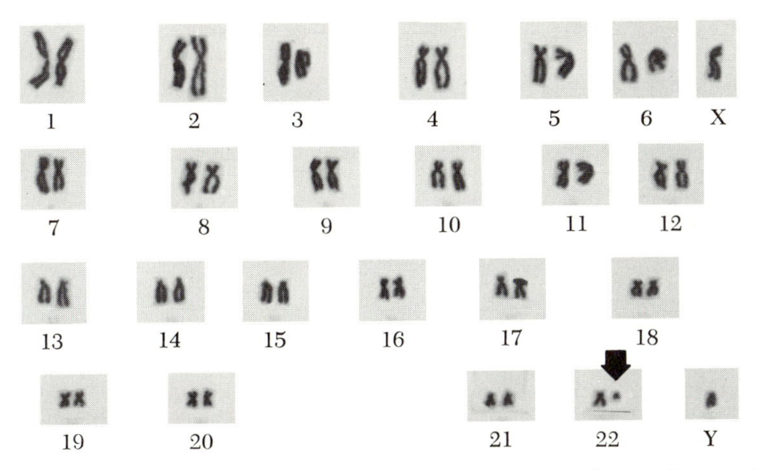

Figure 2-9 Karyotype depicting the presence of one abnormal chromosome, chromosome 22 (arrow), which contains a terminal deletion and is called the Philadelphia chromosome. It is found in patients with chronic myelogenous leukemia. The correlation between this particular chromosomal anomaly and this disease is very high.

missing DNA in this aberrant chromosome (called the Philadelphia chromosome) is less than 1% of the total DNA of the cell. Since one normal homologous chromosome is still present, presumably one copy of each of the missing genes is still available to the cells. Nevertheless, such cells occurring in the bone marrow are always accompanied by malignant disease, and the relationship between the presence of the disease and the presence of this single abnormal chromosome is virtually one to one. Therefore, at the chromosomal level it appears essential to have exactly two members present, either one or three producing a pathological condition.

In contrast to this behavior, many human genetic diseases such as galactosemia and phenylketonuria are caused by a recessive single-gene mutation, where the homozygous condition of the gene defect is required for the disease to be expressed. No chromosomal change is detectable in such patients. The parents of such affected persons are heterozygous with respect to the critical gene and are clinically normal. Therefore, the conclusion seems necessary that while single-gene heterozygosity is often easily tolerated by man, the presence of extra or missing chromosomal regions of a size great enough to be recognized by light microscopy almost always involves severe consequences to the individual. In other words, chromosomal changes appear to be dominant while single-gene changes are usually recessive in their physiological effects. This fact may mean that there are special kinds of chromosomal regulatory mechanisms that require the presence of both members of each chromosome pair, even though recessive-gene mutations may exist in the heterozygous state without producing significant pathology. The existence of the need for chromosomal balance has been recognized in other organisms like plants, but the situation in mammals appears to be especially critical.

In any mammalian cell, only a small fraction of the total genetic potential for protein synthesis is active at any given time. The question arises as to whether all of the genes of each chromosome are activated as a unitary process or whether different parts of different chromosomes are regulated independently. While a mechanism involving a large part or an entire chromosome may be involved in special situations, such as that of the sex chromosomes (see Section 2-6), the data on translocations among human chromosomes suggest that this picture is too simple to explain autosomal regulation in general.

As mentioned on page 35, simple translocations in human chromosomes occur in two forms. Occasionally part of one chromosome appears to have been broken and reattached to an end of another chromosome. If this is the only change in the karyotype and there is no detectable extra or missing chromosomal material, the patient appears completely normal. Such an anomaly is called a balanced translocation and has been observed to involve a variety of different autosomes. Now if each entire chromosome is activated or inactivated as a unit for transcription and protein synthesis, the translocated regions would become activated at improper times and might result in pathological consequences. Since patients with balanced translocations seem completely normal, it appears probable that activation of autosomes occurs in specific positions extending over small chromosomal regions that are independently controlled by metabolic needs of the cell.

When a person with a balanced translocation reproduces, he may donate to the fetus (1) the normal member of each of the two involved pairs; (2) both abnormal members of each pair; (3) the abnormal deficient chromosome together with the normal member of the other pair; or (4) the abnormal chromosome containing the additional fragment together with the normal member of the other pair (Figure 2-6). Each of these events has a probability of 1/4. The offspring in each case, respectively, will be (1) completely normal; (2) translocated in balanced fashion like the parent; (3) imbalanced with a full or partially monosomic chromosome; or (4) imbalanced with a full or partially trisomic region. The last two cases will probably exhibit pathology and may even prove lethal before birth. Since even partial monosomies and trisomies produce pathology, whereas heterozygosity of single genes such as that for galactosemia does not, it seems likely that the human chromosomes contain many loci for regulatory genes whose deviation from the normal number can produce serious biochemical imbalance.

2-6 HETEROCHROMATIN, THE BARR BODY, AND THE LYON-RUSSELL HYPOTHESIS

Interphase nuclei often have heterochromatic (i.e., heterogeneously staining) regions. The existence of these is usually taken to indicate the operation of regulatory mechanisms that render specific parts of the chromosomal complement inactive by keeping them in a highly condensed state in which the individual genes are inaccessible for RNA synthesis. The outstanding case where all or at least a large part of a chromosome's activity appears to be controlled as a unit is that where multiple X chromosomes are present in a cell. An interesting picture that has developed from a series of apparently unrelated studies is emerging. The first of these was conducted by Canadian investigators who were studying the influence on experimental animals of various treatments that might be encountered in wartime conditions. Their investigations revealed the presence of an especially dense body, occupying approximately one-tenth of the volume of the nucleus, in an appreciable fraction of the cells of some of the experimental animals. At first, the appearance of these bodies seemed to be related to the treatment, but it soon became evident that even untreated animals sometimes showed similar structures. The investigators were at a loss to understand the pattern which caused some but not all the animals to exhibit this phenomenon until they realized that these bodies occurred only in cells of females. Since the frequency of occurrence of these bodies is approximately 25% in many of the tissues of the female and is completely absent in the male, it is indeed surprising that this discovery was not made until 1949. Typical pictures of these Barr bodies in female cells are shown in Figure 2-10.

Evidence soon accumulated demonstrating that the Barr body is a

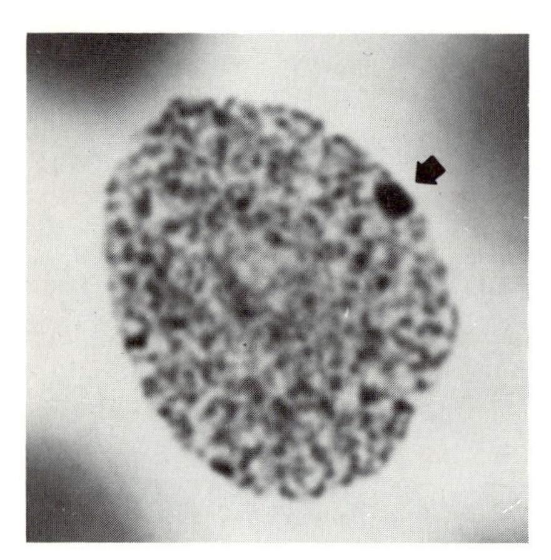

(a)

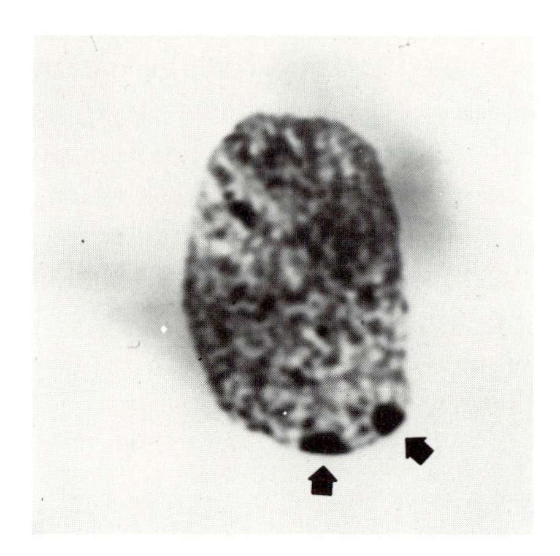

(b)

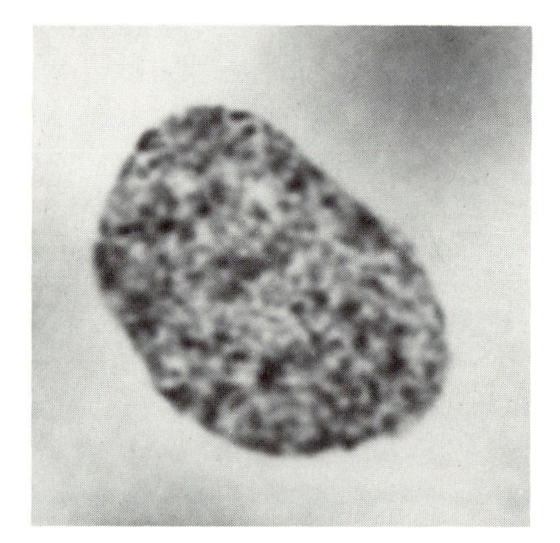

(c)

Figure 2-10 (a) Nucleus of a typical cell from normal female (XX) tissue showing the Barr body, which is formed by condensation of the inactive portion of one of the X chromosomes. (b) Typical cell as in (a), but from an abnormal newborn baby possessing three X chromosomes. Since only one X chromosome can be fully active, this cell contains two Barr bodies. (c) Nucleus of a typical cell from normal male (XY) tissue showing the absence of the Barr body since only one X chromosome is present. A similar picture is obtained in the case of females suffering from Turner's syndrome (XO). Note, however, that the aberrations involving the number of Y chromosomes, such as the XYY condition, would not be detected by this technique, but can be visualized by the fluorescence method.

condensed portion of one of the two X chromosomes of the female somatic cell. It is usually located at the periphery of the nucleus when the cell is stretched out on a glass slide.

While the Barr phenomenon was being elucidated, a theory was advanced independently by investigators in Great Britain and in the United States which furnished a genetic counterpart to the observation of the Canadian scientists. This theory, the Lyon-Russell hypothesis, stated that, in mammalian cells, only one complete X chromosome can be active in protein synthesis at any time. When a cell contains two X chromosomes, a large region on one of them becomes inactive, presumably by assumption of the condensed state which is visible as the Barr body. The decision as to which of the two X chromosomes of any female cell will undergo this inactivation is made early in embryonic development, but when the embryo is already multicellular. The decision is made randomly and is irrevocable so that the clones resulting from each cell have the same pattern of X-chromosome activity as that of the ancestral cell in which the decision was originally made.

This theory made it possible to understand some simple observations such as the mosaic pattern in coat colors in various female mammals. An elegant critical test of this hypothesis was devised by a group in Baltimore using the single-cell plating technique described in Chapter 1. These investigators selected human female subjects who were known to be heterozygous with respect to the gene which is responsible for glucose-6-phosphate dehydrogenase (G6PD), an enzyme coded for by a gene on the X chromosome. The status of these subjects with respect to this enzyme could easily be ascertained on the basis of their having given birth to several males in whom the enzyme was completely absent. Skin samples were taken from such women, single-cell suspensions were made, and the cells were inoculated in petri dishes containing standard growth medium. The developing clones were then individually picked and tested for their enzyme content. A clear prediction of the results to be expected can be made on the basis of the Lyon-Russell hypothesis (Figure 2-11). If both X chromosomes in every cell are metabolically active, the cells of every clone should show the same enzyme activity (which might be half that of a normal homozygous female and the same as that for a normal male cell). The Lyon-Russell hypothesis, however, would predict that some clones would show no activity in any of their cells. These would

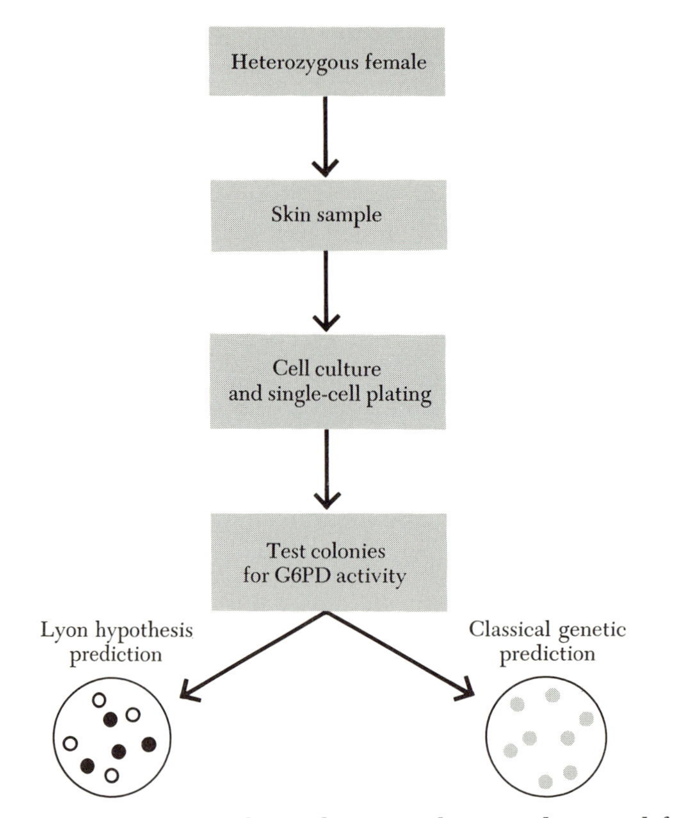

Figure 2-11 *Scheme depicting the procedure used for testing of the Lyon-Russell hypothesis. In the final plates, approximately half of the colonies exhibited full enzyme activity while the remainder showed none. Since this is the distribution to be expected if only one X chromosome is active in any given cell, the validity of the Lyon-Russell hypothesis was confirmed.*

be the clones in which the X chromosome containing the defective gene was the active one. Other clones in which the other X chromosome is active would be expected to show full enzyme activity. The result completely confirmed the expectations of the Lyon-Russell hypothesis. Approximately half of the clones showed no enzyme activity whatever; the remainder had the full complement of G6PD activity.

The confirmation of the validity of the Lyon-Russell hypothesis indicates the great complexity of X-chromosome action in higher organisms. Every female is a mosaic of clones in each of which only one X chromosome is functional. This behavior pattern is also found among other multicellular organisms, such as insects and crustaceans.

The existence of large numbers of sex-linked genes, mutation of which produces disease almost exclusively in males (e.g., hemophilia, color blindness, favism, etc.; see Appendix IV) demonstrates that the possession of a second X chromosome actually does confer genetic advantage on the female. Thus, the mosaicism affords opportunity for genes on both chromosomes to be expressed in a given tissue, even though in any cell only part of a second X chromosome can be active.

The question arises as to whether a selection process might operate in these cells. Thus, in female cells in which one X chromosome is defective, is there opportunity for a selection process to operate that would increase the number of cells with the normal chromosome active? In the case of single-gene defects known to occur on the X chromosome, such as the glucose-6-phosphate dehydrogenase deficiency, experiments appear to be consistent with the absence of selection of cells with one or the other X chromosome active. Thus, in erythrocyte precursor cells, where lack of this enzyme results in defective red blood cells that are extremely sensitive to lysis, and therefore where selection for the normal X chromosome would be of distinct advantage, the numbers of the cells containing an active normal X chromosome and those containing an active defective X chromosome is on the average approximately equal. As noted above, a similar situation exists in skin cells.

The demonstration of the uniqueness of the second X chromosome in mammalian cells raises questions about its behavior in DNA replication. Experiments using radioautographs with tritiated thymidine revealed that, in female cells, one of the two X chromosomes is the last to be replicated (Figure 2-12). Similarly, in male cells, it is the Y chromosome which is the last to be replicated. The lateness of replication of the second X chromosome in mammalian cells is a highly definitive parameter and can be used to identify the X chromosome in cases where translocation or other conditions make its identification doubtful. A rough order also exists in the replication of some of the other chromosomes, although this order is not as definite as is that of the second X chromosome. This ordering in the replication process demonstrates the existence of an information transfer between the chromosomes of the mammalian cell. Thus, if replication of the second X chromosome cannot be completed until all the other chromosomes have been duplicated, some mechanism must exist for transmitting the initiation stimulus to the late-replicating chromosome. The mechanism of this interchromosomal communication remains one of the fascinating unsolved problems of the mammalian cell.

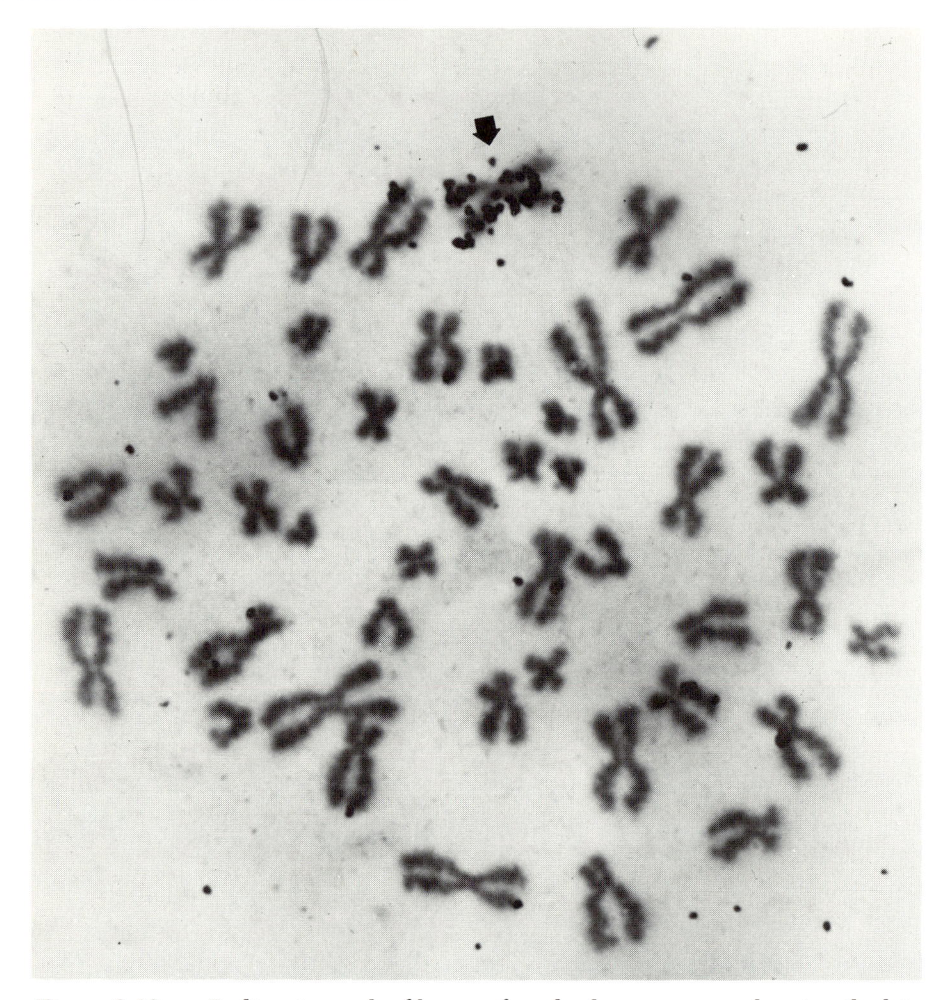

Figure 2-12 Radioautograph of human female chromosomes, showing the late replication of one of the two X chromosomes. H^3-thymidine and colcemide were added simultaneously to a culture of randomly dividing cells, and samples were taken at various intervals to be scored for labeled mitotic figures. The first cells to arrive in mitosis were those which had already completed replication of their DNA and thus were unlabeled. Approximately 2 hours after drug addition, the chromosome pattern pictured above appeared. Since the length of the interval between completion of DNA synthesis and mitosis is about 2 hours (see Chapter 6), the labeled cell was one which was just finishing DNA synthesis when the tritium was added. The only label appears in one of the X chromosomes, indicating that it is the last to be replicated. The other chromosomes, having essentially completed their duplication before the tritium was added, are not labeled. The late-replicating X chromosome is the one which condenses to form the Barr body.

2-7 THE SEX CHROMOSOMES AND SEXUAL DIFFERENTIATION

Until 1959, there was only one animal in which the mechanism of chromosomal sex determination was reasonably well understood, that being Drosophila, which was accordingly taken as a model for animal species in general. However, in Drosophila, the Y chromosome has very few genes. Thus, while two X chromosomes are required for normal female character, a single X chromosome with no Y present (indicated as XO) is a male in Drosophila.

In 1959, human patients with the disease Turner's syndrome were studied and found to possess an XO chromosomal configuration. Such patients are females rather than males, but they are incomplete females because they usually do not undergo sexual maturation at puberty. Therefore, since an XO human is different from a normal XY male, the Y chromosome in man must carry a number of important genes. Patients with the XO condition also often exhibit shortness of stature, lesions of the aorta, a peculiar angle of attachment of the forearms at the elbow, and a characteristic webbing of the skin in the neck. Since these manifestations are absent in normal males and females, it is likely that certain genes are common to the X and Y chromosomes and affect other developmental processes not directly related to primary or secondary sex characteristics.

The conclusions derived from study of the Turner's syndrome conditions in man are borne out by study of the disease Klinefelter's syndrome. Persons with this disease have the XXY condition, that is, they possess both the male and female chromosomal complement. As might be expected, these individuals display some of the sexual characteristics of both males and females, such as prominent breasts and male genitalia. Such persons, like those with Turner's syndrome, are sterile, but unlike the latter are often mentally retarded and emotionally disturbed.

Other anomalous human situations involving multiple X or Y chromosomes and characteristic pathologies have been described. One of these in particular deserves mention. A survey of chromosomal constitution among the inmates of prisons in Scotland revealed that if one considers only the population of males who are more than 6 feet tall and whose prison records indicate a propensity toward crimes of violence, the frequency of occurrence of the sex chromosomal constitution XYY may be as high as 25%. In a normal male population, the incidence of this chromosomal anomaly is less than 0.5%. Confirmatory studies in American prisons have revealed similar relationships. This association of a personality trait with a specific genetic constitution is one of the first such demonstrations in man. It raises fundamental questions about the relationships between human genotype and behavioral patterns, which can have profound implications in the fields of law, sociology, and personality development, as well as genetic biochemistry.

Existence of the Barr body furnishes an exceedingly important means for carrying out epidemiological investigations on the incidence of sex-chromosome anomalies in humans. For example, a normal female and male will display one and no Barr bodies, respectively. However, cells from a Klinefelter's patient (XXY) will show a Barr body, even though the apparent phenotype of the newborn is that of a normal male. Similarly, cells with an XXX chromosomal complement will exhibit two Barr bodies, while those from a female patient with Turner's syndrome (XO) have none (Figure 2-10). In newborns, a rapid, precise procedure can conveniently be applied to cells of the amniotic membrane which arise from fetal tissue and hence have the same chromosomal constitution as the embryo. While cells of most female tissues reveal the presence of the Barr body in approximately 20 to 35% of the cell population, virtually 100% of the cells of the amnion of a female fetus display this marker. Hence, even conditions in which only a small percentage of cells show abnormality can readily be detected. Therefore, if the chromosomal sex of the cells of the amniotic membrane differs from the phenotypic sex of the newborn baby, an anomaly of the sex chromosomes is demonstrated which can be substantiated by complete chromosomal analysis.

The anomalies found in the sex chromosomes far outnumber those in any single autosome pair, a situation that probably reflects the greater ability of the developing embryo to tolerate imbalance in the former without producing death. Future studies should unlock much information about chromosomal and gene influences on sexual development and behavioral patterns in man. The demonstration already available that the Y chromosome appears to make important contributions to "aggressive" behavior itself constitutes an important demonstration that differences in such behavior between the sexes in man may not be attributed purely to cultural influences.

2-8 CHROMOSOMES AND CANCER

It has been pointed out that the cells of all normal human tissues have the same chromosomal constitution. The cancerous condition, however, can be accompanied by a tremendous variety of alterations in karyotype. Changes in both number and structure of individual chromosomes have been reported in various types of cancer. Although a few malignancies have been described, in which no recognizable departure from the normal karyotype is visible, cancer cells more often display karyotypes in which gross aneuploidy and an increase in chromosomal number, often approaching that of the triploid condition, occurs. As yet, there is no clear demonstration as to whether the malignant change induces aneuploidy or vice versa. However, the Philadelphia chromosome (Figure 2-9), which is characterized by a deletion in one of the chromosomes of pair 22, represents a specific association between a given malignancy and a particular chromosomal change.

The need for euploid chromosomal balance described in Section 2-5 might well underlie at least some transformations to malignancy. When this balanced condition is upset early in fetal life, death or developmental anomaly may result. It seems likely, therefore, that the production of a chromosomal aberration upsetting the normal balance in some somatic cells of the mature organism might well produce a cell incapable of obeying the usual signals for stopping or decelerating its rate of multiplication and so result in a cancer.

These considerations are supported by the fact that many, if not most, of the agents known to cause chromosomal aberrations in mammalian cells are also recognized as active carcinogens. In general, these agents may be physical (e.g., ionizing radiation), chemical (e.g., nitrosomethylurea), or biological (e.g., certain RNA and DNA viruses). Some actions of representative agents will be discussed in Chapters 3 through 5.

2-9 POSSIBILITIES FOR CONTROL OF CHROMOSOMAL DISEASES IN MAN

Relatively few fields of science have seen so rapid a bridging between the fundamental discoveries and their application to human problems. Within four years of the initial discovery that the human chromosome number is 46, the enormous role of chromosomal aberrations in human disease had been demonstrated, important medical enigmas were clarified, and the prevention of several kinds of disease became possible. For example, it had long been known that the probability of producing a baby with Down's syndrome is markedly increased in mothers over 35 years of age. However, an appreciable number of affected babies also are born to young mothers. It was found that Down's syndrome patients may display either of two different kinds of chromosomal aberration: the extra chromosomal material of pair 21 responsible for Down's syndrome can arise either from trisomy of chromosome 21 or from the presence of an unbalanced translocation involving attachment of a piece of chromosome 21 to some other chromosome. But whereas trisomy 21 itself occurs only through non-disjunction, a rare phenomenon that becomes more probable with increasing maternal age, the occurrence of a translocation type of Down's syndrome is independent of the maternal age and reveals an inheritance pattern as discussed on page 40 (also see Figure 2-6). With the growth of understanding about these situations, it became possible to offer more definitive genetic counseling to families concerned about the possibility of having children with this disease.

An even more powerful tool has now become available with the development of the procedure called amniocentesis, which involves sampling of the amniotic fluid at periods as early as the 14th week of human pregnancy, and genetic examination of the fetal cells contained therein. In particular, one can make a chromosomal examination of such cells and thus determine whether or not an aberration will be present. If the defect is present, the pregnancy can be safely terminated; if not, the woman can continue her pregnancy free from the anxiety caused by the possibility of having a child with Down's syndrome. In some medical centers, amniocentesis is now offered on a routine basis to pregnant women over the age of 35. Because such women have a much higher frequency of producing trisomy 21 than do younger mothers, an appreciable number of such defective births can be prevented by this means. It has been estimated that the United States spends over one billion dollars a year in caring for patients with Down's syndrome, and this monetary cost is dwarfed by the human cost to the affected persons and their families. Amniotic cells can also be tested for at least some single-gene defects, so that these too can be eliminated by termination of pregnancy of defective fetuses.

While the technique of amniocentesis is extraordinarily powerful in yielding definitive information about the presence of a variety of genetic defects in a developing embryo, its enormous potentiality would not be realized without the social acceptance and permissive legislation necessary to take advantage of these new advances. When chromosome analysis first became possible, a campaign was undertaken to educate legislators as to the definitiveness of the diagnosis of developmental defects due to chromosomal aberration and the enormity of the human and financial cost to the family and to society of bearing and rearing a mentally defective individual. Colorado was the first state in this country to pass legislation making it possible to carry out abortion in cases where there was clear indication that birth, if allowed to occur, would result in a seriously defective individual. As of the time of this writing, 14 other states have changed their laws so as to permit more humane practices in this field. As a result of the combination of circumstances involving advances in scientific understanding, clear and effective presentation of the meaning of these before responsible citizens' groups, and the consequent passage of new laws permitting exploitation of these developments for humane purposes, the possibility of avoiding genetic birth defects has been materially increased. When pregnancy occurs in a family in which there is reason to suspect the possibility of chromosomal aberration, the presence or absence of such a defect can be determined, so that the pregnancy can be safely terminated if an abnormality does exist. Procedures of this kind have become routine in the larger medical centers where the necessary enlightened legislation has been secured. These developments illustrate the growing need for society to realize that a maximally sound genetic constitution must be provided to human offspring. This necessity becomes particularly critical as the need to limit human populations becomes more acute. The tragedies of over-population are compounded when babies continue to be born with severe handicaps such as those of Down's syndrome. The scientist must now assume a new responsibility for explaining to the community the new scientific developments in genetics and their implications for human affairs.

Studies on the factors that produce chromosomal aberrations in cells of man and other animals are under way in several laboratories, although at present only partial hints have emerged about these vital processes. It has been demonstrated that agents such as colchicine, high concentrations of which prevent assembly of the spindle, can in much lower concentration cause chromosomal nondisjunction. Presumably these operate by damaging one or a few spindle fibers so as to prevent equal distribution of the specific chromosomal members between the two daughter cells. Several viruses have been shown to produce extensive chromosome breakage and perhaps also nondisjunction in mammalian cells. An intriguing aspect of the incidence of chromosomal abnormalities in human newborns comes from a study of the time distribution of such birth defects. Instead of following a random distribution, chromosomal aberrations in the newborns of Denver hospitals tended to be markedly clustered in time, with the periods of highest incidence in each year occurring during the months of May through October, corresponding to conceptions occurring during the fall and winter. The reason for this time distribution is as yet unknown.

Amniocentesis makes it possible, even for families possessing an increased risk of chromosomal aberration, to have only normal children, since the abnormal conceptions can be terminated. This constitutes an enormous new power and there is little doubt but that the scope of such power will be greatly increased in the future. In biology, as in physics, the growth of scientific power can equally well be a blessing or a curse to humanity, depending on how it is utilized. Each new stage of scientific development requires intensive reassessment of our most fundamental values and the laws designed to protect them. Without constant reexamination and revision, new scientific developments can result in thwarting the very human purposes which society must try most strongly to protect.

2-10 SUMMARY

Understanding of even the most basic properties of mammalian chromosomes was delayed for decades until appropriate techniques could be developed. The discovery in 1956 that the human chromosome number is 46 rather than 48 was the first of an explosive series of developments, including the classification of human chromosomes according to an internationally accepted system, the identification of various types of chromosomal aberrations, and the correlation of many of these abnormalities with long-recognized human disease conditions. At the present time, over 40 disease-producing chromosomal anomalies, most of which are associated with severe developmental defects including mental retardation, have been well documented. The fact that approximately 1% of all live human births have such aberrations is indicative of the enormous health problem involved.

Summary

Chromosomal aberrations fall into two general classes: those due to the presence of an extra or missing chromosome and those that arise as a consequence of abnormal chromosomal breakage and reunion processes. The tendency to participate in such aberrations appears to be randomly distributed among the different chromosomes, although aberrations in some chromosomes are not tolerated by the embryo and thus do not appear in live births.

Chromosomal-balance factors are enormously important in developmental processes in the mammal. Aneuploidy in man is almost always associated with pathology, even when an extra rather than a missing chromatid results. In the mammalian female, one of the two X chromosomes condenses to form a highly compact, largely inactive structure observed as the Barr body. The demonstration that this second X chromosome is the last chromosome to replicate in a multiplying female cell is an indication that information transfer occurs between the mammalian chromosomes. Similar mechanisms involving chemical exchanges of products between autosomes may well play an important part in development and could explain the large pathological consequences of aneuploidy in mammals. Aneuploidy is often observed in malignant cells, and one type of cancer, chronic myelogenous leukemia, has been identified with the presence of a specific chromosomal aberration.

At present, the following fundamental questions remain to be elucidated in mammalian cell genetics:

(1) How can a general method for locating genes on chromosomes be developed?
(2) What is the mechanism by which specific genes are made active or inactive in accordance with normal developmental processes?
(3) To what extent can such gene regulatory mechanisms be modified or reversed?
(4) By what mechanism does the process of nondisjunction occur and how can it be prevented?

Identification of chromosomal anomalies in man has made possible an impressive beginning in the prevention of many kinds of birth defects. These new developments promise to produce enormous new powers for control of human reproduction in a variety of ways. Broad understanding of these possibilities and enlightened revision of laws and social practices are essential if the utilization of these new powers is to benefit mankind.

REFERENCES

General

Dupraw, E. J. "Cell and Molecular Biology," Academic, New York, 1969.

German, J. Studying human chromosomes today, *Amer. Scientist* **58**, 182 (1970).

Ham, R. G., and T. T. Puck. Quantitative colonial growth of isolated mammalian cells, *Methods Enzymol.* **5**, 90, 1962.

Ohno, S. "Sex Chromosomes and Sex-linked Genes," Springer, New York, 1967.

Puck, T. T., and A. Robinson. Some perspectives in human cytogenetics, in "The Biological Basis of Pediatric Practice," vol. 2, edited by R. E. Cooke, McGraw-Hill, New York, 1968.

Yunis, J. J. Human chromosomes in disease, in "Human Chromosome Methodology," edited by J. J. Yunis, Academic, New York, 1965.

Selected papers

Barr, M. L., and E. G. Bertram. A morphological distinction between neurones of the male and female, and the behavior of the nucleolar satellite during accelerated nucleoprotein synthesis, *Nature* **163**, 676 (1949).

Book, J. A., J. Lejeune, A. Levan, E. H. Y. Chu, C. E. Ford, M. Fracarro, D. G. Harnden, T. C. Esu, D. A. Hungerford, P. A. Jacobs, S. Makino, T. T. Puck, A. Robinson, J. H. Tjio, D. G. Catcheside, H. J. Muller, and C. Stern. A proposed standard system of nomenclature of human mitotic chromosomes, J. Amer. Med. Assoc. **174**, 159 (1960).

Boyle, J. A., K. O. Raivio, K. H. Astrin, J. D. Schulman, M. L. Grax, J. E. Seegmiller, and C. B. Jacobsen. Lesch-Nyhan syndrome: Preventive control by prenatal diagnosis, *Science* **169**, 688 (1970).

Caspersson, T., L. Zech, C. Johansson, and E. J. Modest. Identification of human chromosomes by DNA-binding fluorescent agents, *Chromosoma* **30**, 215 (1970).

Davidson, R. G., H. M. Nitowsky, and B. Childs. Demonstration of two populations of cells in the human female heterozygous for glucose-6-phosphate dehydrogenase variants, *Proc. Natl. Acad. Sci.* **50**, 481 (1963).

DuPraw, E. J. "DNA and Chromosomes," Holt, New York, 1970.

Lejeune, J., M. Gautier, and R. Turpin. Etudes des chromosomes somatiques de neuf enfants mongoliens, *Compt. Rend.* **284**, 1721 (1959).

Lyon, M. F. Sex chromatin and gene action in the mammalian X chromosome, *Amer. J. Human Genetics* **14**, 135 (1962).

Russell, L. B. Genetics of mammalian sex chromosomes, *Science* **133**, 1795 (1961).

Tjio, J. H., and A. Levan. The chromosome number of man, *Hereditas* **42**, 1 (1956).
Tjio, J. H., and T. T. Puck. The somatic chromosomes of man, *Proc. Natl. Acad. Sci.* **44**, 1229 (1958).

CHAPTER 3

Single-Gene Mutagenesis

The techniques of molecular biology permitted detailed analysis of the biochemical events involved in the normal and pathological metabolism of microorganisms. Their application required the availability of a variety of different markers, such as virus and drug resistance and specific nutritional deficiencies, and of selection systems permitting isolation of the desired mutants.

In applying these methods of analysis to the somatic mammalian cell, additional complexities were encountered, which required special developments. First, these are diploid cells. Unlike most microorganisms studied, they do not have a haploid phase of their life cycle, which would permit ready recognition of recessive mutations. Further, a selection system for recessive mutations, permitting isolation of the defective mutants, is necessary. Problems also exist due to chromosomal instability, which under certain conditions can lead to rapid shifts of karyotype in a mammalian cell culture. Finally, the huge numbers of mammalian cell genes increase enormously the amount of work needed to gain detailed understanding of the genetic potentialities of the mammalian cell.

The problem of suitable genetic markers has been especially difficult. When mammalian cells are exposed to lethal doses of drugs or viruses, the majority of the cell population may be killed and apparently resistant clones will appear. On subculturing, however, most of these will often be found to be sensitive, their apparent resistance having been a physiological rather than a genetic property. One approach to a solution of this problem has been the use of multiple-step changes, obtained by successive treatment of cell cultures with progressively increasing doses of lethal agents. Ultimately, clones of high and permanent resistance have been established in this way. However, while such cell markers are useful for some kinds of experiments, the fact that a number of independent mutations may have contributed to the final behavior of the cell makes it difficult to relate the changed phenotype to changes in specific genes.

Another approach that has proved very useful has utilized a spontaneous mutant of a mouse cell (the L fibroblast) which lacks the enzyme thymidine kinase. The properties of this mutant and some of the ways in which it has been used are mentioned in Chapter 4. A third system, involving the Chinese hamster ovary cell, offers several unique advantages for study of these problems and is discussed in some detail in the following section.

3-1 DESCRIPTION OF THE CHINESE HAMSTER CELL SYSTEM

The Chinese hamster cell was selected for these studies because its small number of chromosomes (11 instead of 23 pairs as in man) affords a great simplification. In addition, stable, long-term cultures can readily be established from cells taken from a variety of tissues, and experiments are rapid because the generation time of these cells is only 12 hours, approximately half the generation time of human, rat, and mouse cells. One particular Chinese hamster cell, which had originated from a culture of the ovary, was found to be hypodiploid. Thus, even recessive mutations, if they occur in monosomic regions, should be readily demonstrable.

Single-step mutations would appear to offer the best opportunity for establishing relationships between specific genes and their associated products. Mutations resulting in specific nutritional deficiencies were sought in the hope that these would permit more direct biochemical analysis. The first step in the development of such a system is the achievement of a completely defined basal medium containing the minimal number of nutrients necessary to promote growth of single cells into colonies with high efficiency. Such a medium (F12D) was developed for the Chinese hamster ovary cell (CHO), and is shown in Table 1-1. When mutants are produced that cannot grow in this basal medium, their growth may be tested in the enriched medium, F12, which contains nine additional metabolites: alanine, glycine, glutamic acid, aspartic acid, vitamin B_{12}, lipoic acid, inositol, thymidine, and hypoxanthine. It is then a simple task to determine the specific nutritional requirement gained by a mutant which no longer can grow in F12D but can grow in F12.

Next, a method of selection of the nutritional mutants is required. However, since the great majority of mutations produce loss of function and are recessive, their isolation from the parental population is difficult. Thus, if one cell in a million has mutated to a condition requiring an additional nutritional supplement, both this mutant and the remainder of the population will grow equally well in the presence of the supplement, but this mutant alone will not grow in its absence. Therefore, the problem of isolating this rare mutant cell from the more self-sufficient members of the population present in overwhelmingly larger numbers is difficult. This problem has been solved in bacterial genetics by the "penicillin technique," which utilizes the fact that addition of penicillin to cells in an unsupplemented medium selectively destroys the multiplying wild-type cells. Thus it is possible to enrich the population in the proportion of cells requiring nutritional supplement.

A method similar in principle to this technique was developed for mammalian cells and is shown schematically in Figure 3-1. It depends upon the following facts:

(1) Cells in nutritionally complete medium can synthesize DNA while those unable to carry on protein synthesis cannot.
(2) Bromodeoxyuridine (BUdR), a biologically active analog of thymidine, when added to the medium, becomes incorporated specifically into DNA.
(3) When cells which have incorporated BUdR into their DNA are illuminated with long ultraviolet and with visible light (sometimes called near-visible light), these cells are rapidly destroyed, but normal cells are unchanged.

Consequently, cell populations containing suspected nutritionally deficient mutants are placed in a medium which will support growth only of the wild type, and BUdR is added. Thereafter, the entire cell population is illuminated with near-visible light, which can be obtained from standard daylight fluorescent lamps, so that those cells able to multiply in the minimal medium are selectively killed. The medium then is enriched with various metabolites, and those cells that had not been able to multiply because of a nutritional deficiency now grow out to form colonies. These colonies are picked, and their

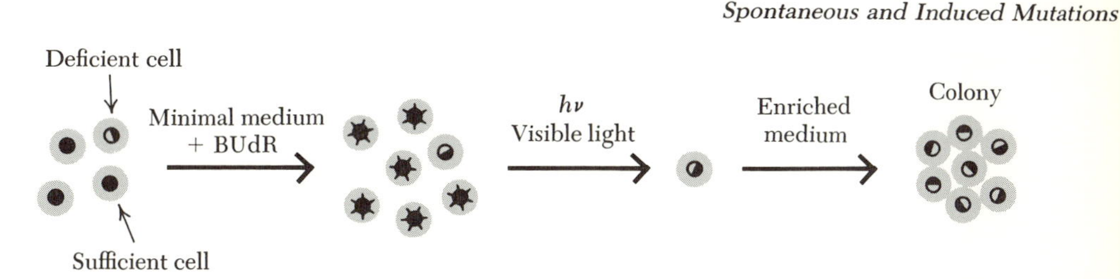

Figure 3-1 Schematic representation of the BUdR-visible-light technique for isolation of nutritionally deficient mutant clones. The mixed cell population is exposed to BUdR in a deficient medium in which only the normal, wild-type cells can grow. These alone incorporate BUdR into their DNA and are killed on subsequent exposure to a standard fluorescent lamp. The medium is then changed to a composition lacking BUdR but enriched with nutrilites that permit the deficient cells to grow up into colonies.

specific nutritional deficiences identified. This method has been found to be so reproducible in its operation that it can be used not only to obtain deficient mutants but also to quantitatively measure their frequency of occurrence in a large population of the sufficient cells.

3-2 SPONTANEOUS AND INDUCED MUTATIONS

One spontaneous auxotrophic mutation involving a requirement for proline was found in the previously described hypodiploid cell, which had originated in a Chinese hamster ovary and had been grown in tissue culture with stable karyotype for many years. Using it as the parental cell, a variety of other nutritional mutations were induced and then isolated by means of the BUdR-visible-light selection method.

The mutagens tested in the first series of studies included both physical and chemical agents as shown in Table 3-1. After treatment with agents known to be effective mutagens in bacterial systems, large numbers of nutritionally deficient mutants were produced. Mutations were readily obtained for the requirements for glycine, hypoxanthine, inositol, thymidine, and various combinations of these metabolites, but no requirement for the other five molecules of the enrichment solution has so far been observed despite the screening of very large numbers of cells. One possible explanation for this difference in behavior may be that the latter five metabolites are synthesized through the agencies of genes diploid in the CHO cell, while the others are contained in the hemizygous portion. In Table 3-2 are shown the numbers of the various kinds of mutants isolated after treatment with agents used in this study. The cells were treated with the agent for a period of approximately one generation time, in the case of the chemical agents, and the concentration of the chemical agent or the dose of the radiation employed was such as to produce

Table 3-1 Representative Mutagens Tested

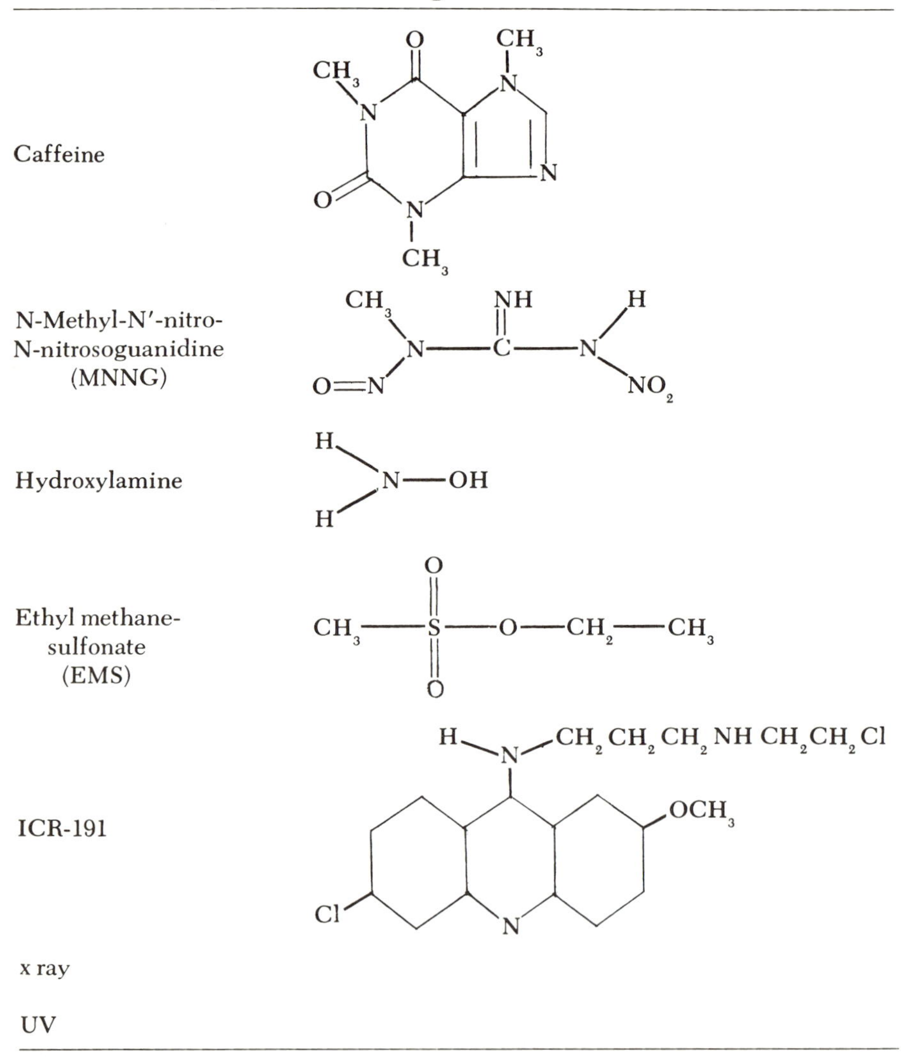

Table 3-2 Auxotrophic Mutants Produced by Treatment with Various Agents (approximately 6 x 10^5 surviving cells were tested for each mutagen, and three times this many cells for the control)

	Number of Mutants Found			
Agent	gly⁻	gly⁻ hyp⁻ thy⁻	ino⁻	hyp⁻
None (control)	0	0	0	0
EMS	20	1	0	0
MNNG	4	10	1	0
ICR-191	4	0	0	4
Hydroxylamine	0	0	0	0
Caffeine	0	0	0	0
UV	1	9	0	0
x ray	2	0	0	0

cell survival in the range of 12 to 78%. The first point to be noted is the absence of mutations in the untreated cultures. Tests revealed the spontaneous frequency of revertants of these deficient mutants to be usually less than 10^{-8}. Thus the parental cell and the mutants derived from it are genetically stable. The original clone and its mutant subclone also display a satisfactorily constant hypodiploid constitution with a modal number of 20. Therefore, these mutants are capable of being used for genetic tests with high resolving power. Figure 3-2 demonstrates the behavior of a glycine-requiring mutant (gly^-), when plated in the presence and absence of glycine, and its response to various concentrations of that metabolite.

3-3 COMPLEMENTATION ANALYSIS ON MAMMALIAN CELLS *IN VITRO*

The existence of single-gene mutations that have very low spontaneous reversion rates and that produce all-or-none growth of single cells makes possible rapid and convenient complementation analysis utilizing the techniques of cell fusion, as developed by English and French workers (see Chapter 4). Their experiments demonstrated cytoplasmic and nuclear fusion resulting either when mammalian cells are treated with Sendai virus, whose own reproduction and capacity for cell killing has been destroyed by a previous exposure to

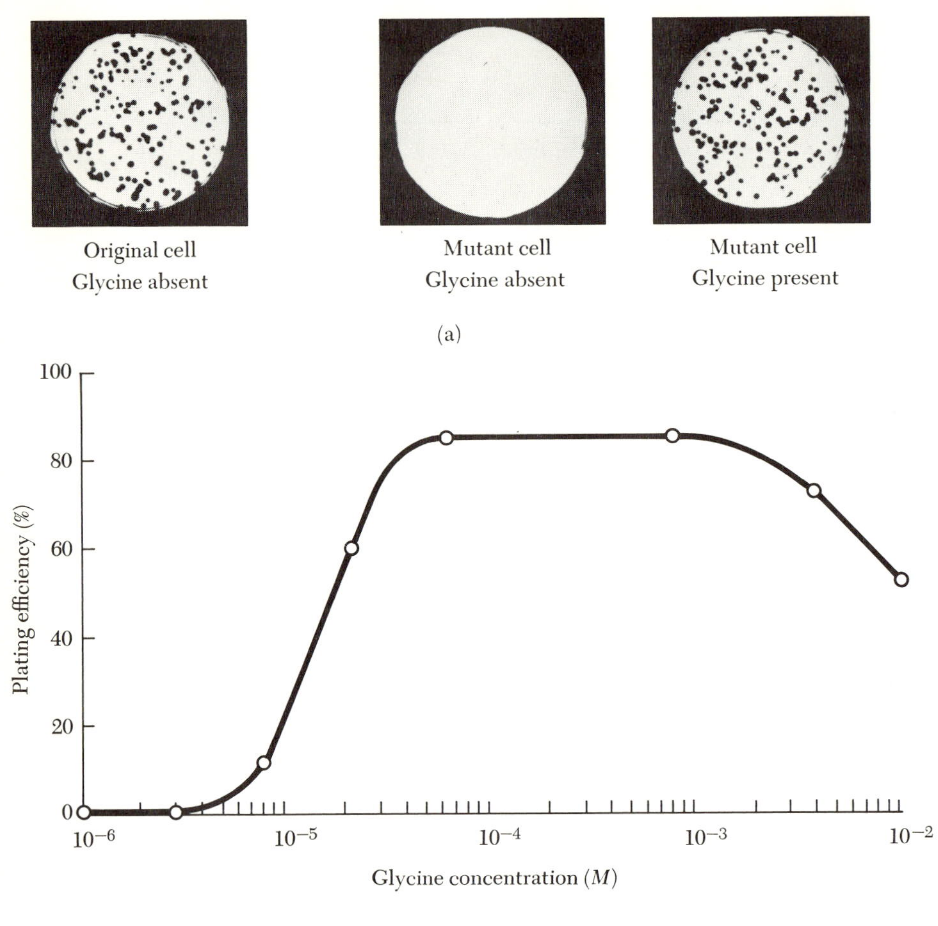

Figure 3-2 *The response of the gly⁻ mutant to glycine. (a) Colonial growth in the presence and absence of glycine, (b) response of the mutant to various concentrations of glycine.*

ultraviolet irradiation (Figure 3-3), or through a process of spontaneous cell fusion. While the molecular mechanism of cell fusion is still obscure, this procedure has become an extremely useful tool. The fused cells first form a single cell with multiple nuclei. These nuclei then fuse to form a single large nucleus, which, at first at least, contains all of the chromosomes of each of the component cells. Both processes yield essentially similar results, but the use of irradiated Sendai or other viruses increases the frequency of cell fusion by

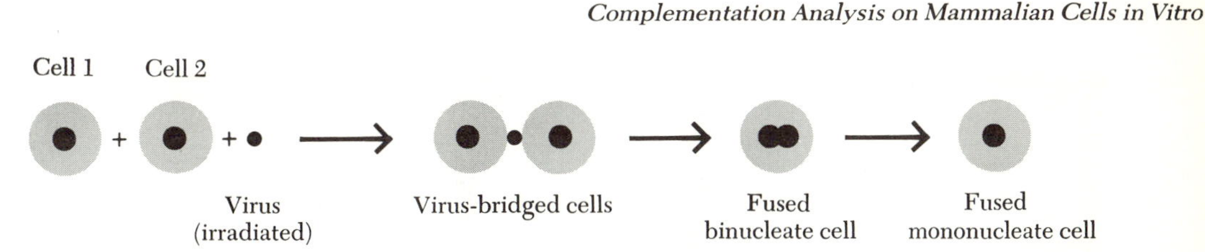

Figure 3-3 Schematic diagram illustrating the process of virus-induced cell fusion. Cells will also fuse spontaneously in the absence of virus, although the frequency of this process is approximately 1000-fold lower than the former.

approximately 1000-fold. In either case, however, the result of a fusion experiment is to bring together in a single cell the chromosomes and the cytoplasmic components of two separate cells. The behavior of the resulting cell then offers means for drawing conclusions concerning the interactions of the contributing elements (Figure 3-4).

The simplest kind of complementation analysis involves fusion between cells that differ from each other in single-gene mutations only. In such cases the assumption seems fairly safe that the cytoplasmic elements are essentially constant, and the behavior of the heterokaryon can be interpreted in terms of the differences in action of the given nuclear genes. One of the immediate applications of the technique is to decide whether given mutations are recessive or dominant.

Consider the fusion of a glycine-requiring (gly^-) and a hypoxanthine-requiring (hyp^-) mutant of the CHO culture. In the standard procedure, 5×10^5 cells of each type are placed together in growth medium from which both glycine and hypoxanthine have been omitted. Ultraviolet-inactivated Sendai virus is added to the mixture and, after incubation first at 4°C and then at 37°C for a time sufficient to produce fusion in approximately 2% of the cells, aliquots of the cell mixture are plated in media containing or lacking the two critical nutrients. The results obtainable in such an experiment are illustrated in Figure 3-5. It is seen that whereas each mutant alone exhibits no growth whatever unless its specific nutritional requirement is supplied, 1 to 2% of the cells plated from a fusion mixture have grown to colonies in the absence of both of these nutrients. When these colonies are picked, they breed true and yield stable clones that exhibit high plating efficiencies and growth rate in the absence of the two nutrients required by the parental cell. Thus, the mutations to glycine

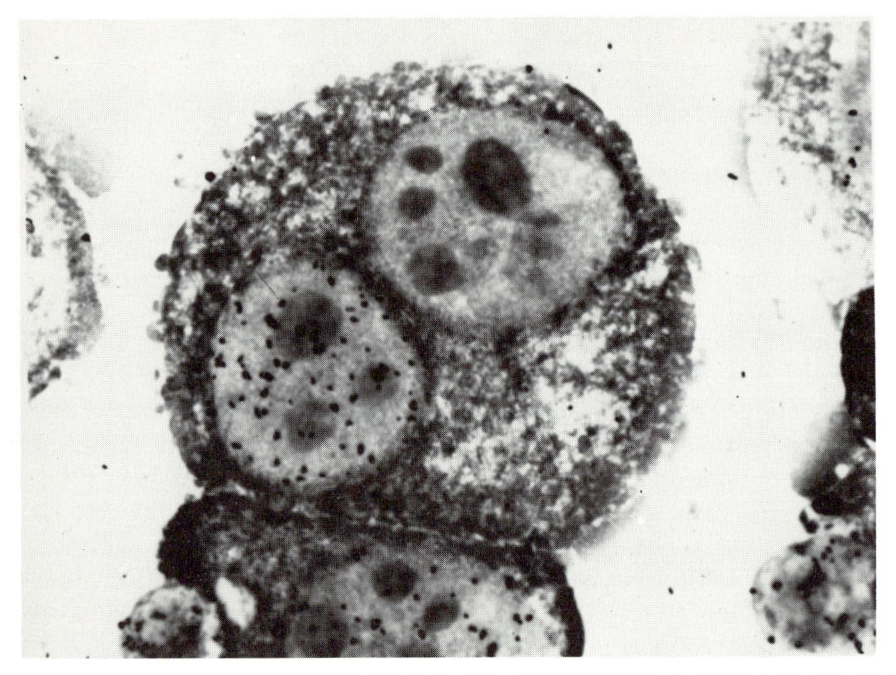

Figure 3-4 Photomicrograph of a hybrid binucleate cell formed by the fusion of two CHO cells. The black-speckled nucleus is that of a cell which was exposed to H^3-thymidine before fusion; the light nucleus is from one which was untreated. These two nuclei will fuse to produce a cell with a single nucleus containing the chromosomes of both parental cells. Hybrid cells of this type thus offer a valuable system for many kinds of genetic experiments.

and hypoxanthine requirements, respectively, are recessive, since the fused cells show neither deficiency. Presumably, recessive mutations could be due to loss of a structural gene, while dominant mutations, as in bacteria, could result from the production of an inhibitory substance, thus preventing expression of the active gene contributed by the other fusion partner.

When the karyotypes of the fused cells are examined, they are seen to contain chromosome numbers approximating twice that of the parental cells. With the passage of time, there is a slight tendency to lose chromosomes, but eventually the chromosome number stabilizes at a value approximately 10% less than the sum of the chromosome number of the parental cells (Figure 3-6).

Complementation study in fused cells also permits a variety of other kinds of genetic analysis. For example, in an effort to obtain nutritional mutants of the Chinese hamster cell, it was found that mutants requiring glycine were obtained in greater frequency than those requiring any other metabolite. Two different kinds of glycine-requiring mutants were found. The one required glycine alone to restore completely the growth capacity of the cells; the other required glycine plus a purine (hypoxanthine) and a pyrimidine (thymidine) in order to restore growth. The question then arises as to whether all of these glycine-requiring mutants have undergone mutation at the same gene or whether a variety of different genes could mutate so as to introduce a glycine requirement for cell multiplication.

Cell-fusion experiments can readily answer such a question. Mutants deficient only for glycine were fused with mutants containing the triple deficiency for glycine, thymidine, and hypoxanthine. If both cells contained a mutation in the same gene required for glycine synthesis, the product of such fusions should still require glycine for growth. On the contrary, such experiments revealed that fused cells completely lost the requirement for glycine (as well as that for hypoxanthine and thymidine). Therefore the glycine requirement in these two mutants must stem from mutation in different genes.

Analysis was then carried out to determine whether all of the mutants requiring glycine alone had become defective in the same gene. Since fusion between cells with the same defect never produces

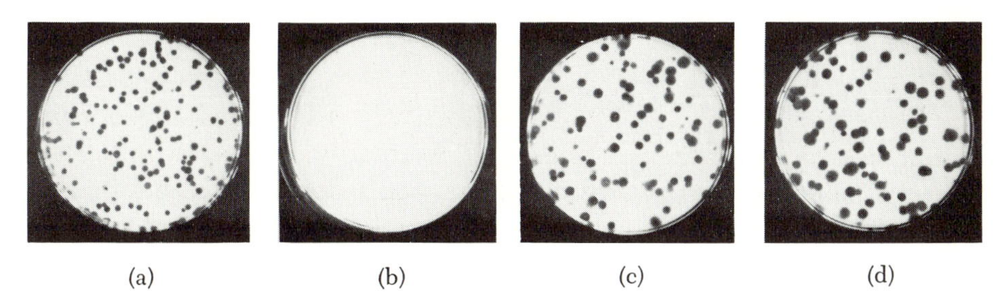

(a) (b) (c) (d)

Figure 3-5 *Demonstration of use of nutritional markers to study cell fusion. (a), (b) Growth of the gly$^-$ mutant in the presence and absence of glycine, respectively. A hyp$^-$ mutant exhibits the same contrast in the presence and absence of hypoxanthine. (c) Growth in complete medium of the hybrid cell formed from fusion of the gly$^-$ and hyp$^-$ mutants. (d) Growth of the same hybrid in medium lacking both glycine and hypoxanthine. Thus each mutant has contributed to the hybrid cell functional genes which were lacking in the other. Such experiments demonstrate that each of the defective genes was recessive. These nutritional markers make rapid fusion experimentation possible, since every colony developing in the deficient medium must have arisen from a hybrid cell.*

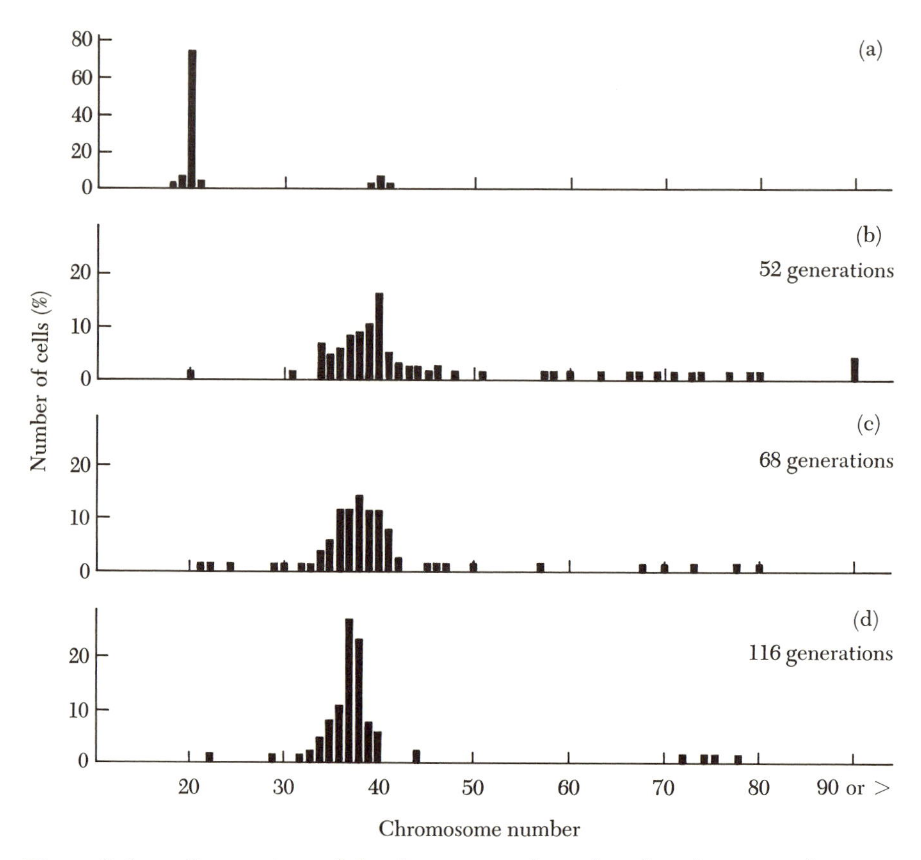

Figure 3-6 Comparison of the chromosomal number distribution in the parental and hybrid cells. (a) Chromosomal distribution of either parental cell. (b)-(d) Chromosomal distribution of a typical hybrid cell clone, as determined at various periods of growth after the fusion. The initial counts approximate the tetraploid number, but also contain appreciable numbers of more polyploid forms. With the passage of time, the modal number decreases by 5-10%, and most of the forms with chromosome numbers greater than 40 disappear. By the 116th generation, a concomitant increase in plating efficiency from about 50% to 70-80% has occurred, presumably because of the population drift to a distribution of forms with more stable chromosomal constitutions.

glycine-independent hybrids, while fusion of cells with different defects always does, various combinations of the 13 different mutants were fused with each other, two at a time, and the resulting hybrid cells tested for growth in glycine-free medium. Such experiments revealed the 13 mutants to comprise four mutually exclusive complementation groups (Table 3-3). Thus, mutation at any of four different genetic loci is able to introduce a requirement for glycine into these cells. All four types of mutation appear to be recessive since they can be complemented by fusion with cells in a class other than their own.

Table 3-3 Demonstration of the Existence of Four Classes of gly⁻ Mutants on the Basis of Complementation Through Cell Fusion (the plus sign indicates that the resulting hybrid can grow in medium lacking glycine)

	gly_A^-	gly_B^-	gly_C^-	gly_D^-
gly_A^-	−	+	+	+
gly_B^-	+	−	+	+
gly_C^-	+	+	−	+
gly_D^-	+	+	+	−

3-4 IDENTIFICATION OF THE METABOLIC BLOCKS

The metabolic scheme leading to proline synthesis is shown in Figure 3-7 and is reasonably similar in bacterial and mammalian cells. The fact that glutamic acid, even in large concentrations, will not produce growth of the deficient mutant indicates that the proline-deficient mutant is blocked at a point at or after the conversion of glutamic acid to its γ-semialdehyde. Moreover the γ-semialdehyde will support growth in a fashion equivalent to that of proline. Therefore the blocked reaction is that shown in Figure 3-7. (Under special circumstances, one can demonstrate a small amount of glutamic γ-semialdehyde production from ornithine by reversal of the process shown in the figure. However, under standard test conditions, this reaction does not occur to an appreciable extent.)

Similar biochemical analysis can be performed on each mutant so produced, and is a necessary part of the identification of the genetic changes produced. The case of the gly⁻ mutants is of special interest. Glycine is formed in these cells from serine by means of the enzyme serine hydroxymethylase and the cofactor tetrahydrofolate, as shown in the following reaction:

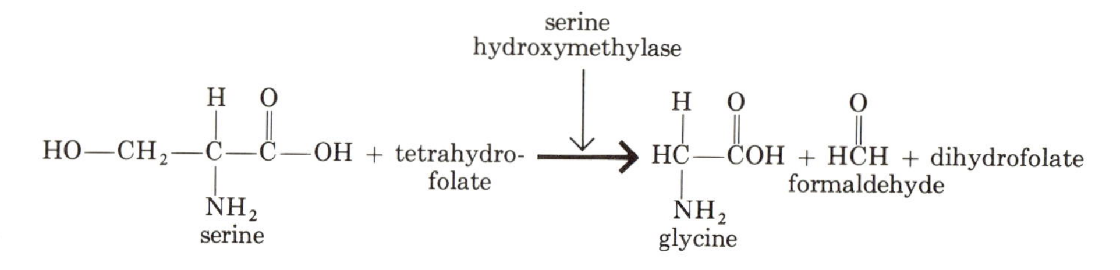

The metabolic blocks leading to the gly⁻ deficiency must be different in each of the four complementation classes because each class involves a different gene. Therefore, biochemical analysis was carried out on cells of each of the four groups, in an attempt to characterize them biochemically as well as genetically. Experiment demonstrated that Class A mutants lack the enzyme serine hydroxymethylase. This enzyme is present in all of the remaining mutants, but is unable to function because of a deficiency of tetrahydrofolic acid in these cells. At the time of this writing, the exact site of the metabolic block in each of the B, C, and D mutants has not yet been identified, but definite differences in their biochemical behavior have been found, as illustrated in Table 3-4. Thus it is now possible to identify the class of any of these glycine-requiring mutants, either by the genetic test of complementation or by biochemical tests. The availability of such mutants and genetic operations illustrates how far progress has been made in adapting to studies on mammalian cells the powerful methods that were first developed for microorganisms, such as *Escherichia coli.* It would appear that means are now available for genetic analysis with reasonably high resolving power, and that elucidation of genetic-biochemical relationships may be possible for large numbers of genes connected with mammalian cell multiplication.

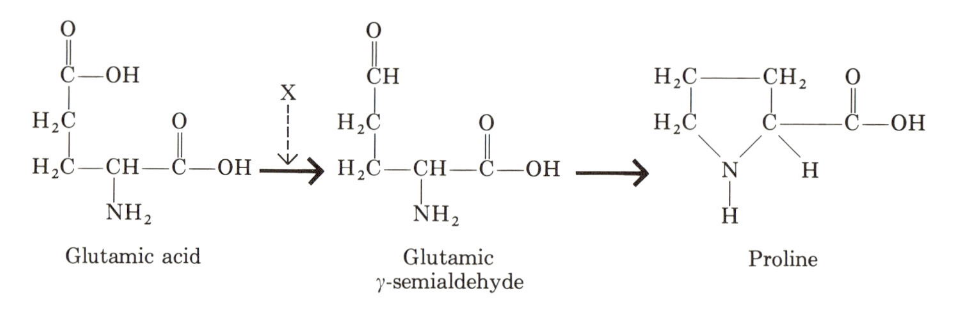

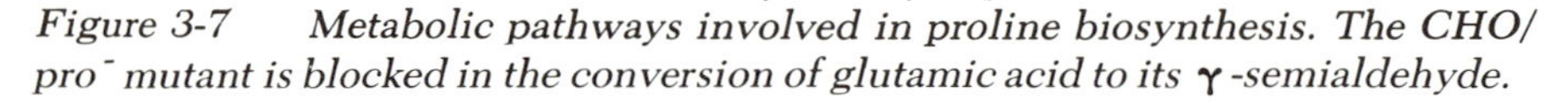
Figure 3-7 Metabolic pathways involved in proline biosynthesis. The CHO/pro⁻ mutant is blocked in the conversion of glutamic acid to its γ-semialdehyde.

Table 3-4 Summary of Distinguishing Properties of the Various gly⁻ Complementation Classes

Complementation Class	Presence of Serine Hydroxymethylase	Restitution of Growth by Folinic Acid* Alone	Restitution of Growth by Folinic Acid plus Trace of Glycine
A	0	0	0
B	+	Complete	Complete
C	+	0	Partial
D	+	0	0

*Folinic acid ≡ N^5-formyl tetrahydrofolic acid.

3-5 QUANTITATION OF MUTAGENESIS IN MAMMALIAN CELLS

One would like to compare quantitatively the degree of mutagenesis produced by treatment of mammalian cells with various agents. Since most mutagens also kill cells, one would also like to know the relationship of mutation production to cell killing for any treatment. The first step in the process, then, involves measurement of the survival curves for each agent. Such survival curves for a variety of mutagenic agents are presented in Figure 3-8. It is obvious that the curves vary widely. It is therefore difficult to select a concentration for different chemical compounds at which their mutagenic action could be compared. Moreover, it is also difficult to compare the action of chemical agents with those of physical agents, such as x irradiation and ultraviolet light.

However, the survival curves shown in Figure 3-8 exhibit a similar pattern: An initial shoulder may or may not exist in the curve, but once the curve begins to fall it follows a straight line when plotted as a semilogarithmic graph.[1] It then becomes convenient to utilize a quantity, D^0, defined as the mean lethal dose.[2] When the data of Figure 3-8 are recalculated in terms of the dose divided by D^0 the series of curves resulting is that shown in Figure 3-9. Now all of the curves

[1]This behavior is predicted by a random-hit mathematical model. This model assumes that each agent exerts a damaging action on the cell by means of an independent event at a particular locus and that the probability of such an event increases in a linear fashion as the dose of the agent is increased.

[2]D^0 is defined as the concentration or dose needed to reduce the number of viable cells by 63%, as measured in the linear portion of the curve. D/D^0 corresponds to the average number of lethal events suffered by the cell, if the random-hit model is applicable.

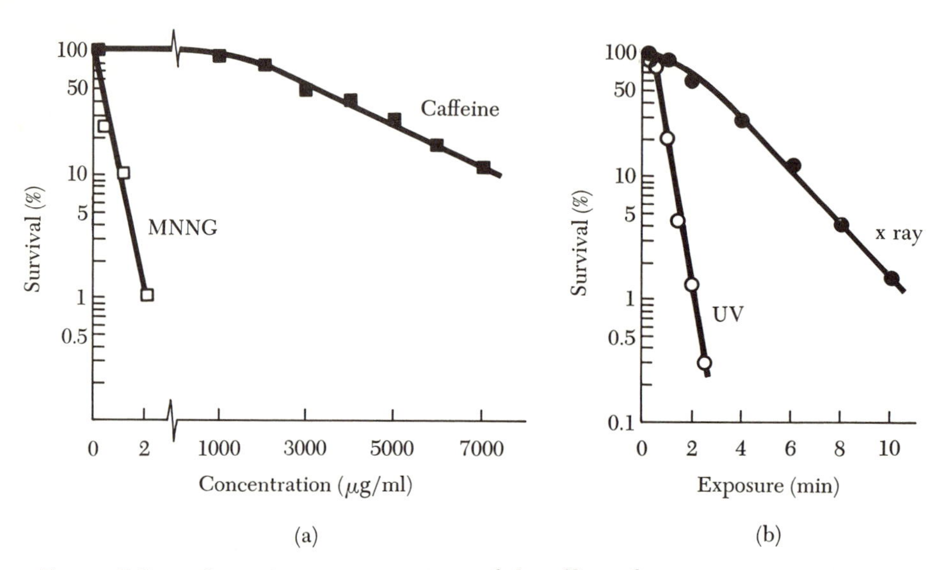

Figure 3-8 ***Quantitative comparison of the effect of various mutagenic agents on cell survival. The wide range of effective concentrations and limiting slopes make difficult selection of a concentration at which the corresponding mutagenic activities could be compared. In addition, it is difficult to compare meaningfully the actions of chemical agents (a) with those of physical agents (b). A solution to these difficulties appears in the data of Figure 3-9.***

have the same limiting slope but are displaced from each other by an amount reflecting the difference in their respective initial shoulders. Such curves make it very convenient to compare cell-killing behavior of different agents. As long as the limiting slope is constant, the presumption is that the killing mechanism is constant. At a point where this slope might change, one would have reason for suspecting that some new parameter has entered to affect the mechanism of cell killing. This behavior then makes it convenient to select the D^0 value for each agent as the standard concentration or dose at which its mutagenic effects are to be compared. The D^0 values for several mutagenic agents are shown in Table 3-5.

Mutagenic agents operating on mammalian cells will produce reversions of auxotrophic mutations as well as forward mutations of prototrophs to auxotrophs. The latter kind of mutagenesis is more difficult to score, but it is more frequent and more useful in trying to screen agents that may represent health hazards because of their mutagenesis. As shown in Figure 3-10, reversion of an already existent mutation can only be brought about by an agent capable of reversing the effect of the specific lesion previously introduced into the gene. A very powerful mutagen might appear to have little or no activity by such a test. In the measurement of forward mutagenesis, however, an agent that is able to introduce a defect anywhere in the gene's structure so as to cause loss of its biological activity will become manifest.

In Table 3-6, the single-gene mutagenesis and the effectiveness in producing chromosomal breaks is summarized for a variety of different agents. In addition to chromosome breaks, one can also score

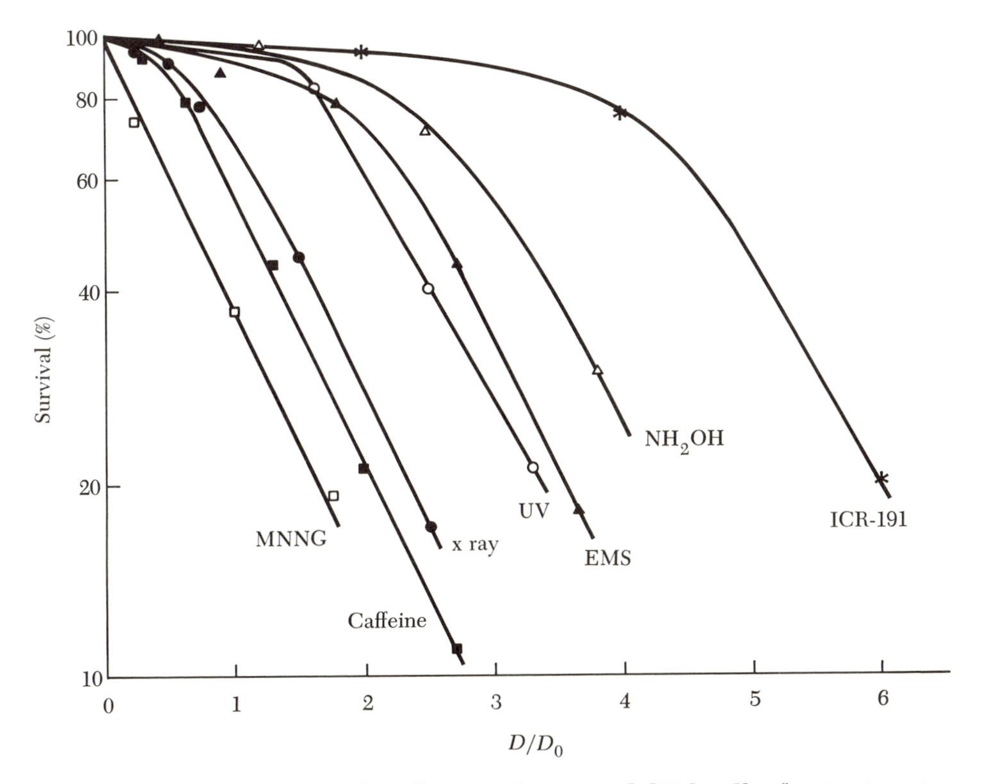

Figure 3-9 Modified single-cell survival curves of CHO cells after treatment with the various mutagens indicated. In order to alleviate the difficulty discussed in Figure 3-8, the dose of each agent is expressed in units of its D^0 value as listed in Table 3-5. The fact that all the curves exhibit the same limiting slope implies that the data fit the hit model of killing actions. This model assumes that a target region exists within the cell in which the agent's lethal action can be exerted; and that the probability per unit dose or concentration of exerting a lethal hit within the target is constant, after the initial shoulder concentration is exceeded.

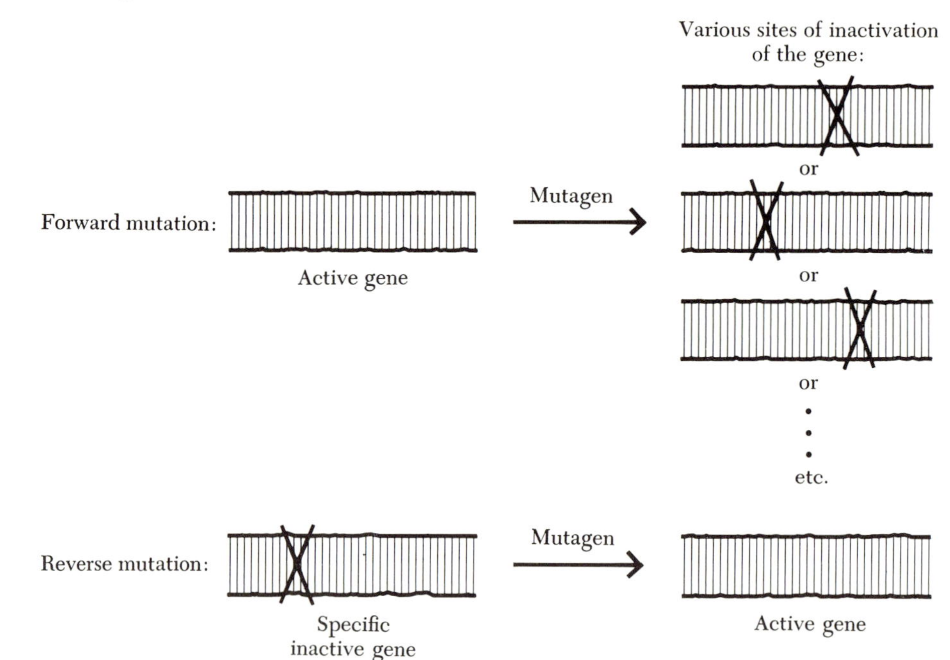

Figure 3-10 Schematic illustration of the fact that forward mutations can be produced by a defect any place in the gene's structure, while reversions occur only by action at a specific inactive locus. Mutagenesis in the forward direction, therefore, is more effective in the screening of drugs and other agents for possible mutagenic action.

Table 3-5 Comparison of D^0 Values for a Variety of Mutagenic Agents

Mutagen	D^0	
EMS	110.0	γ/ml
MNNG	0.44	γ/ml
ICR-191	0.25	γ/ml
NH_2OH	25	γ/ml
Caffeine	3000	γ/ml
UV	69	ergs/mm^2
x ray	200	rads

chromosomal exchanges which represent the rearrangement of chromosome fragments produced by the breaks. The scoring of these is shown in Figure 3-11. Thus, the action of three different kinds of

Table 3-6 Comparison of Mutation Frequency and Chromatid Breaks Produced by Mutagenic Agents

Mutagen	*Forward Mutation Frequency per Gene per 100 cells per D^0*	*Average Chromatid Breaks per 100 Cells per D^0*
None	0	0
EMS	0.013	6.3
MNNG	0.0035	9.7
ICR-191	0.00033	0.3
NH_2OH	0	2.0
Caffeine	0	8.7
UV	0.0015	27.3
x ray	0.00072	42.7

agents has been elucidated: those, such as ethylmethane sulfonate, that produce both chromosomal breaks and rearrangements, as well as single-gene mutations; agents, such as caffeine, that in certain concentration ranges produce chromosomal changes almost exclusively; and agents, such as ICR-191, that can be used to produce only gene mutations. The availability of these different kinds of agents should make possible study of the genetic effects of both single-gene and chromosomal mutations.

3-6 REVERSION OF MUTATIONS TO PROTOTROPHY

The single-gene mutations described in the preceding paragraphs show a range of spontaneous reversion, varying from 10^{-6} for the proline mutant to about 10^{-8} or less for the other mutations.

The reversibility of two gly⁻ A mutants, which had been induced by different agents, was examined in order to identify the mutation type. One mutant, which was produced by x ray, showed neither spontaneous nor mutagen-induced reversion, even though 10^8 or 10^9 cells were tested. Therefore, this mutation probably is the result of a deletion. Another gly⁻ A mutant, produced by ethylmethane sulfonate, was found to be reverted readily by ethylmethane sulfonate and a variety of other alkylating agents and so would appear to be due to a base change. These methods then permit analysis of the molecular basis of the mutations produced exactly as in bacteria.

The question then arises as to the genetic nature of requirements

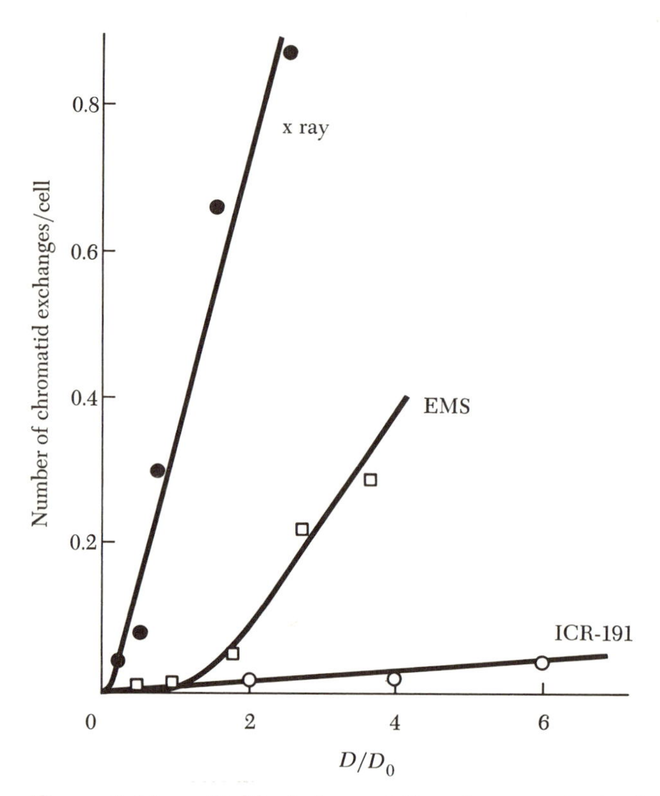

Figure 3-11 Yield of chromatid exchanges obtained from exposure of CHO cells to various kinds of mutagenic agents. Dosage of each agent is expressed in terms of its D^0 value (see Table 3-5). X ray exhibits the highest and ICR-191 the lowest yield of chromosomal exchanges in relation to the killing dose. Other agents tested (e.g., EMS, MNNG, UV, caffeine, and hydroxylamine) yielded intermediate values.

for a growth factor in a presumably diploid cell, since the probability of two independent mutations at each of the two sister genes on homologous chromosomes is too small to be seriously considered. While a single dominant mutation could produce a nutritional requirement, this possibility has been ruled out by complementation studies. Examination of the karyotype of the CHO cell revealed it to contain only 21 chromosomes (Figure 3-12). While, unfortunately, some chromosomal rearrangement has also occurred, producing nine members of the karyotype with structures different from the normal, there appears to be little doubt that some chromosomal material is missing, so that presumably the genes on these missing regions are present in the hemizygous state. Therefore, a spontaneous mutation in any one of the genes of this hemizygous region would produce a phenotypic change even if the mutation were recessive.

These considerations suggest an approach for dealing with one of the most serious problems of genetic-biochemical analysis in diploid cells. If one can produce a variety of cells monosomic in different chromosomes, the genes on those chromosomes will be hemizygous and therefore will readily produce mutations that will result in a change in phenotype. When monosomic cells are produced with otherwise normal karyotypes, it should be possible by this method to identify the chromosomal site of the genes in which mutations become readily demonstrable. When clones from the CHO culture were examined, one was found with only 19 chromosomes, and this has been selected for more intense genetic study. If these nutritional mutants do indeed arise from a mutation in the monosomic region of

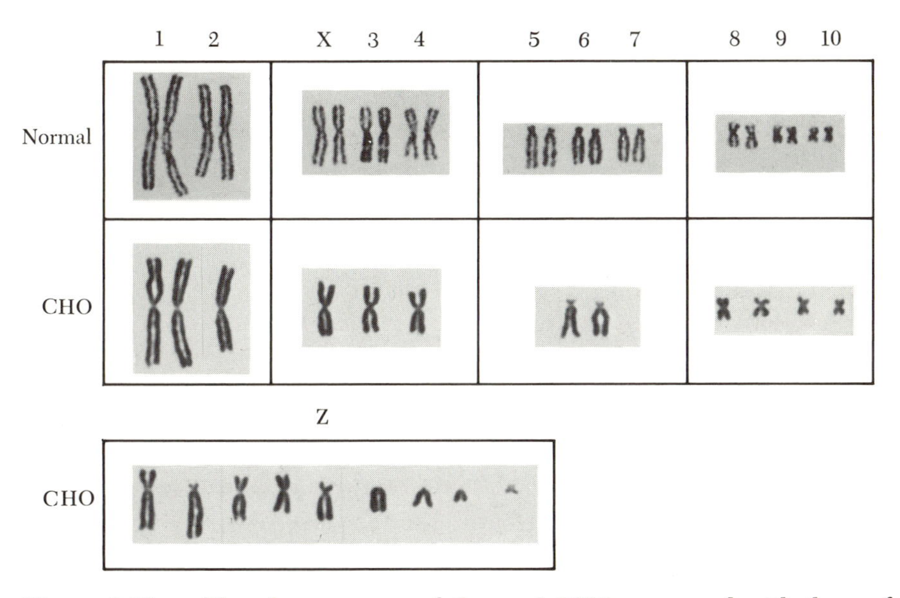

Figure 3-12 The chromosomes of the pro⁻ CHO compared with those of a normal Chinese hamster cell. Some chromosomal rearrangements have occurred in CHO, producing nine chromosomes which cannot be identified precisely and thus have been placed in group Z. Quantitative examination of the two karyotypes, however, indicates that some chromosomal material has been lost in the CHO cell, and presumably this material contained one of the two genes necessary for proline synthesis, so that a mutation in the other produced the deficient phenotype.

the chromosome material, it follows that a revertant cell should be in the hemizygous condition, and it should be possible to demonstrate differences between the behavior of these cells and homozygous cells with a full complement of chromosomes.

Experiments were carried out in an effort to determine whether the revertant from the proline-requiring cell would show the expected differences in behavior when compared with a normal cell. For the latter purpose, a culture of a Chinese hamster lung cell, which has a modal chromosomal number of 23, was selected. When tested for their growth response to proline, cells of both cultures grew with 100% plating efficiency in the presence or absence of proline. However, while the fully diploid cell shows no response whatever if proline is added to the medium, the revertant grows slowly in the absence of proline but its rate is markedly accelerated when proline is added to the medium (Figure 3-13).

Similarly, the cells were compared in their ability to synthesize radioactive proline from the radioactive precursor, glutamic acid. The chromosomally complete form produces a large conversion of glutamic acid to proline; the proline-deficient form has none, the amount of radioactive proline formed being essentially background. Four different revertants selected at random each produce an amount of conversion which is approximately one-half that of the chromosomally complete cell. These data, then, suggest that the revertant is indeed in the hemizygous state.

3-7 HUMAN IMPLICATIONS

Human genetic diseases fall into two groups: chromosomal aberrations (Appendix II) and gene mutations (Appendix IV). Of the latter group, some of the gene defects underlying the pathological conditions have already been identified, but many others are still unknown. Moreover, the genetic basis for many human developmental processes and their resultant traits are obscure. The ability to produce single-gene mutations in mammalian cells *in vitro* opens up many possibilities for genetic studies in man. These techniques will undoubtedly lead to the identification of genes that are as yet unrecognized. In addition, they should yield understanding of the specific mutations underlying human genetic diseases, allow more accurate evaluation of threats due to mutation by environmental agents, and help clarify the role of mutational processes in mammalian and human evolution.

CHO/pro^{+} revertant

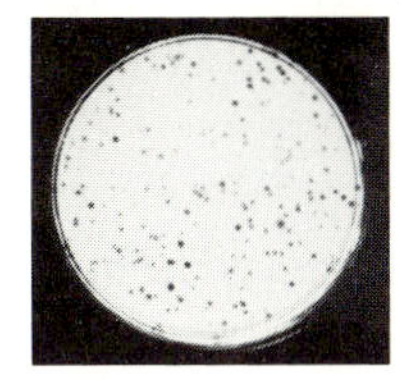

Proline: 0

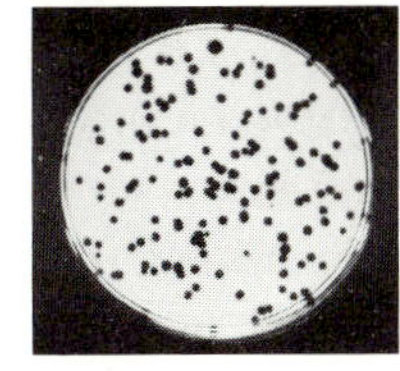

3×10^{-4} *M*

Figure 3-13 Demonstration that although the CHO/pro^{+} revertant can grow with high plating efficiency in the absence of proline, maximum rate of growth is not obtained until proline is added. This would be as expected if the revertant contained only one copy of the proline genes. The normal cell, which presumably contains two copies of the proline gene, grows equally rapidly with or without the addition of proline.

For several decades, human geneticists have been disturbed about the possibility that the progress of medicine, by keeping alive and allowing the reproduction of individuals with genetic defects, might bring about accumulation of defective genes in the human population. The results discussed in this and the preceding chapter conceivably might offer a possible antidote to this pessimistic outlook. Twenty to twenty-five single-gene defects are now detectable by amniocentesis. This procedure makes possible avoidance of homozygous and, theoretically, at least, even heterozygous defective genes by a process of fetal selection that enables couples to avoid transmitting their own genetic defects to their children. As the number of potentially pathological conditions that can be diagnosed in this manner increases, it may well be that by this method, which involves rational choice, man may be able to guide his evolution in such a way as to eliminate those genes whose accumulation might otherwise constitute a serious threat.

The experimental results discussed in this chapter have opened up new methods for genetic-biochemical analysis of the action of mammalian genes that affect cell reproduction. Forthcoming developments should provide simple, powerful means for study of mechanisms that control differentiation processes. Methods for detection and isolation of clones, mutated with respect to differentiation processes, and delineation of the genetic biochemistry associated with these processes can be expected to furnish the next large advance in mammalian cell biology. It is conceivable that understanding of the differentiation process may permit intervention in the processes of embryonic development. One may be enabled to repair defects or even to improve on normal developmental processes in such a way as to augment human abilities in areas such as increased resistance to infection, improved operation of the nervous system, or increased ability to withstand the ravages of the aging process.

The data of this chapter emphasize the new opportunities that loom ahead for man in understanding his own genetics and genetic biochemistry. It must be borne in mind, however, that the genetic constitution determines only the potentialities of any organism. To achieve fullest fulfillment of human potential, much remains to be learned about how experience can affect human development, both before and after birth. For example, it has already been amply demonstrated in mice that nutritional deprivation during and shortly after gestation produces animals with irreversible damage to their intellectual abilities. New evidence exists that a similar situation may exist in man. Man is undoubtedly the animal most sensitive to the effects of experiential conditioning in modifying the realization of his genetic potential. It is vital to carry on both kinds of studies in man.

3-8 SUMMARY

Elucidation of the biochemical genetics of mammalian cells for a long time remained at a relatively low level as compared with bacteria, largely because the greater complexities of the mammalian cell made clean and well-defined experiments prohibitively difficult. Recent advances in this area, however, promise to extend greatly the kinds of experiments which can be performed.

It is now possible to (1) induce single-gene nutritional mutations by a variety of chemical and physical agents, (2) isolate these mutants from large populations of wild-type cells, and (3) quantitate the frequency of forward and backward mutagenesis, as well as cell killing and chromosomal damage, which often accompany mutagenesis. In addition, complementation analysis between auxotrophic mutants may be used to determine the dominance or recessiveness of a specific mutation and the identity or nonidentity of mutations leading to the same phenotype. By assaying the activities of specific enzymes in the metabolic pathway, the location of the mutation may then be pinpointed.

These developments have enormous practical as well as theoretical value. Routine screening of food additives, drugs, and environmental pollutants for mutagenic action is now possible. Methods for testing chemical and physical agents, with respect to production of mutagenesis and associated processes, may provide valuable insights into the relationship between mutagenesis and carcinogenesis. Finally, delineation of the genetic biochemistry of differentiation processes and their control holds promise, in the future, for repair of certain genetic defects and perhaps also the promotion of certain developmental processes that may improve the health and well being of the human organism as a whole.

REFERENCES

General

Ham, R. G., and T. T. Puck. Quantitative colonial growth of isolated mammalian cells, *Methods Enzymol.* **5**, 90, 1962.

Harris, H. "Cell Fusion," The Dunham Lectures, Harvard, Cambridge, Massachusetts, 1970.

Selected papers

Barski, G., S. Sorieul, and F. Cornefert. Production dans des cultures *in vitro* de deux souches cellulaires en association de cellules de caractère "hybrid," *Compt. Rend.* **251**, 1825 (1960).

Boyle, J. A., K. O. Raivio, K. H. Astrin, J. D. Schulman, M. L. Graf, J. E. Seegmiller, and C. B. Jacobsen. Lesch-Nyhan syndrome: Preventive control by prenatal diagnosis, *Science* **169**, 688 (1970).

Chu, E. H. Y., and H. V. Malling. Mammalian cell genetics II. Chemical induction of specific locus mutations in Chinese hamster cells *in vitro*, *Proc. Natl. Acad. Sci.* **61**, 1306 (1968).

Kao, F. T., L. A. Chasin, and T. T. Puck. Genetics of somatic mammalian cells X. Complementation analysis of glycine-requiring mutants, *Proc. Natl. Acad. Sci.* **64**, 1284 (1969).

Kao, F. T., R. T. Johnson, and T. T. Puck. Genetics of somatic mammalian cells VIII. Complementation analysis on virus-fused Chinese hamster cells with nutritional markers, *Science* **164**, 312 (1969).

Kao, F. T., and T. T. Puck. Genetics of somatic mammalian cells IV. Properties of Chinese hamster cell mutants with respect to the requirement for proline, *Genetics* **55**, 513 (1967).

Kao, F. T., and T. T. Puck. Genetics of somatic mammalian cells VII. Induction and isolation of nutritional mutants in Chinese hamster cells, *Proc. Natl. Acad. Sci.* **60**, 1275 (1968).

Kao, F. T., and T. T. Puck. Genetics of somatic mammalian cells IX. Quantitation of mutagenesis by physical and chemical agents, *J. Cell. Physiol.* **74**, 245 (1969).

Puck, T. T., and F. T. Kao. Genetics of somatic mammalian cells V. Treatment with 5-bromodeoxyuridine and visible light for isolation of nutritionally deficient mutants, *Proc. Natl. Acad. Sci.* **58**, 1227 (1967).

Sorieul, S., and B. Ephrussi. Karyological demonstration of hybridization of mammalian cells *in vitro*, *Nature* **190**, 653 (1961).

Weiss, M. C., and B. Ephrussi. Studies of interspecific (rat x mouse) somatic hybrids. I. Isolation, growth, and evolution of the karyotype, *Genetics* **54**, 1095 (1966).

CHAPTER 4

Genetic Analysis in Mammalian Cells

Linkage analysis in an organism like the mammal involves identification of individual genes, assignment of them to their chromosomes, and establishment of their order and distance apart on each chromosome. These operations have in the past been particularly difficult in man because conventional genetic procedures for determining linkage are difficult to apply in an organism with which one cannot carry out specific genetically illuminating matings. Population study and pedigree analysis, which might conceivably furnish the needed information, are costly to carry out, and the results are attended by much more uncertainty than when large-scale controlled mating experiments can be performed. For this reason, then, the chromosomal distribution and mapping positions of the human autosomal genes have remained almost completely unknown.

To a nongeneticist, the problem of locating the positions of the genes on the 23 pairs of human chromosomes might seem a purely academic pursuit. On the contrary, it is one that has the deepest practical significance for human and medical genetics. When large numbers of the human genes have been mapped it will be possible to know which gene sequences are deficient or supernumerary in patients with monosomies, polysomies, deletions, and translocations. This information should suggest therapeutic or preventative approaches for at least some of these conditions. Moreover, availability of such information should make possible determination, by biochemical analysis, of the exact location along the gene sequence at which translocations or deletions may have occurred. But an even more fundamental development may be expected from such studies: Gene mapping in bacteria made discovery of the operon, the simplest and most fundamental of the genetic regulatory mechanisms, possible. The problems of genetic control in mammalian cells appear to be even more complex than in bacteria. There is little question that many human genetic defects involve aberrations of regulatory mechanisms. It becomes necessary then to be able to map the human genes so as to gain understanding of the nature of gene control in mammals and to apply this knowledge to problems of cellular biochemical control in health and disease.

4-1 PRINCIPLES OF GENE LOCALIZATION

Since it is now possible to grow, recognize, and isolate clonal colonies of mammalian cells, to identify their chromosomes, to label the DNA with a specific marker, such as tritiated thymidine, and to produce hybrid cells by cell fusion, several different approaches to determining the chromosomal distribution of genes would appear to be feasible.

Cell fusion with chromosomal loss. It has been demonstrated that fusion between somatic cells of certain species, for example, man and the mouse, produces hybrid forms which undergo extensive chromosome loss, primarily of one species. Thus, if an animal cell containing a gene deficiency that prohibits growth in the standard medium, is fused with a human cell, the hybrid clones so produced can be isolated after extensive loss of human chromosomes (Figure 4-1). The gene derived from the human cell that compensates for the animal cell mutation is presumably contained on one of the human chromosomes common to all such clones. By means of cytologic examination of a sufficiently large number of such clones, one could then hope to identify the chromosome on which the gene in question is contained. For such a method to be useful on a large scale, it is necessary that (1) large numbers of mutant animal cells be available, (2) the mutations have sufficiently low reversion rates so that every colony is indeed a hybrid, (3) each mutation block the biosynthesis of a specific metabolite, and (4) the complementing chromosomes be readily identifiable.

These principles were used to demonstrate that the gene responsible for the formation of thymidine kinase lies on one of the group E human chromosomes. A mutant of the mouse L fibroblast cell lacking the gene for the enzyme thymidine kinase was used. The major purine

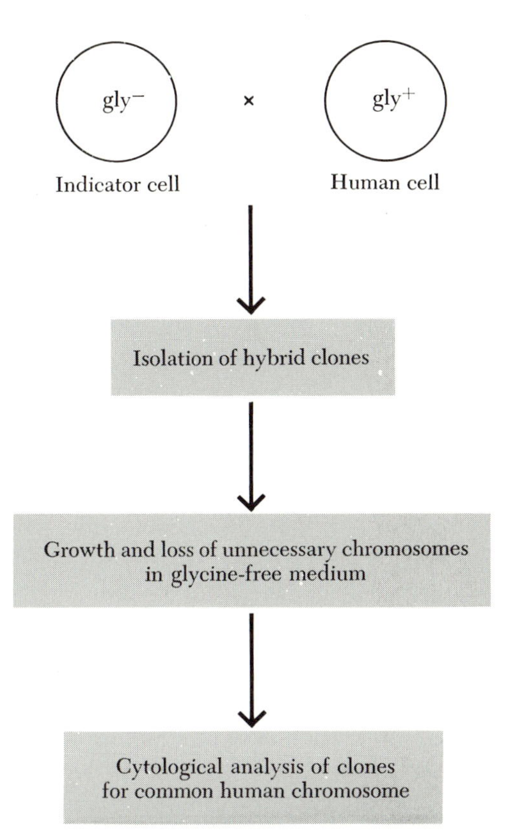

Figure 4-1 *Demonstration of the use of cell-fusion techniques for mapping of genes on chromosomes. The human cell overcomes the gene deficiency of the gly⁻ mutant. Initially all the hybrid cells will grow in the glycine-free medium. However, as more and more of the human chromosomes are lost, only those hybrids that retain the chromosome containing the gly⁺ gene will remain viable. Cytological examination of the viable clones, then, should reveal this chromosome common to all such clones.*

and pyrimidine biosynthetic pathways can be blocked by the drug aminopterin. In its presence, normal cells still can multiply if they make use of auxiliary pathways and if they are furnished with thymidine, hypoxanthine, and the amino acid glycine. However, the mutant cell lacking thymidine kinase cannot utilize exogenous thymidine and therefore is unable to grow in the presence of the so-called "HAT medium," (containing hypoxanthine, aminopterin, and thymidine).

Fusion of this mutant cell with a normal human cell produces a hybrid that can grow in the HAT medium since the human cell furnishes the gene necessary for formation of thymidine kinase. Through study of the distribution of human chromosomes in various hybrids formed between the mouse mutant and human cells, a good correlation was found between retention of human chromosomes of the submetacentric E group and the presence of the thymidine kinase gene in the resulting hybrid.

These experiments revealed the great importance of utilizing the principle of chromosomal loss from cell hybrids to locate genes on the human chromosomes. However, the system utilizing the mouse cell presents certain experimental difficulties. The large number of the mouse L cell chromosomes may introduce some uncertainty in identifying which chromosomes of the mixed karyotype are human and which of these is uniquely common to all the different clones that have been relieved of their gene deficiency. There is also the possibility that over the long periods of time required by these experiments, chromosomal breakage and rearrangement may occur. In this case the problem of chromosomal identification is aggravated. Occasionally when human-mouse hybrids are employed, reduction of the human chromosome number to a low level may not occur, so that it becomes laborious or impossible to identify a single chromosome which is common to all the hybrid clones. The number of well-characterized genetic markers that can be used for these experiments is at present limited in the mouse L cell. Finally, the slow rate of growth of the human-mouse hybrid cells requires relatively long periods for completion of an experiment.

The Chinese hamster cell system described previously offers a number of possible advantages for such studies. The first lies in the rapid growth rate of this cell. In the human-mouse cell hybrid, both parental members have generation times in the neighborhood of 16 to 20 hours. The Chinese hamster cell, however, reproduces much more rapidly. When fusion is carried out between cells of widely different generation times, the growth rate of the slowest member should at first dominate that of the hybrid. However, if chromosomes of the more slowly growing form are lost, the growth rate of the resulting cell approaches that of the faster growing form. Therefore, if a human-Chinese hamster cell hybrid loses Chinese hamster chromosomes, the resulting form should grow slowly. If it loses human chromosomes, it should grow more rapidly. Thus, such a hybridization provides the possibility for automatic selective advantage promoting rapid isolation of hybrid cells that have lost the maximum number of human chromosomes.

As discussed in the preceding chapter, the Chinese hamster cell offers a low chromosome number and a variety of different genetic markers. At this time, ten different markers involving nutritional deficiencies have been produced, with reversion rates sufficiently low to make clean experiments possible, and additional markers appear to be obtainable. Clones with multiple genetic markers can be readily produced so that determination of linkage becomes possible. Thus, a Chinese hamster cell with two auxotrophic deficiencies can be fused with a normal human cell in an appropriate selection medium. Growth and chromosome loss of the resulting hybrids is carried out in the presence of only one of the critical metabolites. Clones are then isolated and tested for the presence of the second nutritional deficiency. If such clones always exhibit the presence of both deficiencies, the presumption is that both genes are carried on the same chromosome (provided that the different clones contain only one human chromosome in common). On the other hand, if a large fraction of the picked clones have lost the nutritional requirement that did not enter into the selection method, one can conclude that the two genes are unlinked (Figure 4-2).

In a typical experiment, double mutants of the Chinese hamster cell that lacked the genes needed for both glycine and proline synthesis (gly^-pro^-) were fused with normal human fibroblasts and hybrids formed. Chromosomal examination of such cells was carried out at various periods after fusion in order to determine the rate of loss of human chromosomes. Figure 4-3 shows the results of chromosomal counts made on such a hybrid culture after one, two, and five weeks of growth. One week after fusion the chromosome number has reached a point close to that of the original Chinese hamster cell. Within one or, at the most, two weeks, the clones achieve a stabilized chromosomal constitution. All of those so far studied have lost the great majority of their human chromosomes and, sometimes, a few of the Chinese hamster chromosomes as well.

In testing for linkage, fusion of normal human fibroblasts and the doubly deficient Chinese hamster mutant, gly^-pro^-, was again carried out, and hybrid cells were grown up in the presence of proline but in the absence of glycine. After two weeks of growth and chromosome loss in this partially deficient medium, hybrid cells were selected and

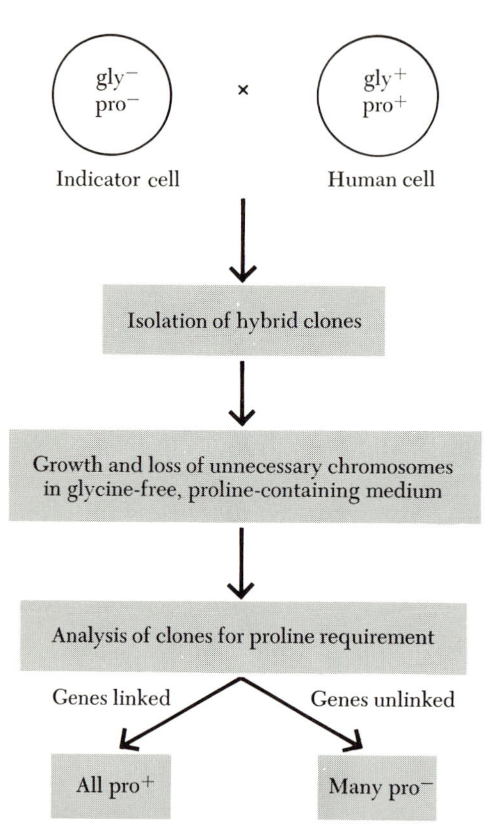

Figure 4-2 Procedure for determination of gene linkage. The process is similar to that in Figure 4-1 except that at least two markers are necessary. As the hybrid cells grow in the glycine-free, proline-containing medium, a selective loss of human chromosomes occurs with the result that only those cells that retain the gly$^+$ human chromosome remain viable. The presence of pro$^+$ in all gly$^+$ clones would strongly indicate the two genes are on the same chromosome. However, the appearance of pro$^-$ would indicate that the pro$^+$ gene had been lost while the gly$^+$ gene was retained. In this case, therefore, the two genes must be unlinked.

their nutritional requirements determined. As expected, all of the clones were able to grow in the absence of glycine. However, the proline requirement of the original Chinese hamster cell was found to be variable, indicating that the two genes studied were unlinked. Thus, one-third of the clones picked were found to require neither glycine nor proline (gly$^+$pro$^+$), while two-thirds required only proline (gly$^+$pro$^-$). Similar experiments demonstrated that the gene for proline synthesis is unlinked to that required for inositol synthesis in the human karyotype.

Figure 4-4 shows the karyotype of the original human parental cell, the karyotype of the CHO subclone employed and the karyotype of the stabilized hybrid requiring proline but not inositol ($ino^{+}pro^{-}$). The stabilized hybrid cell contains only 19 chromosomes, one less than the number in the original Chinese hamster cell. Of this number,

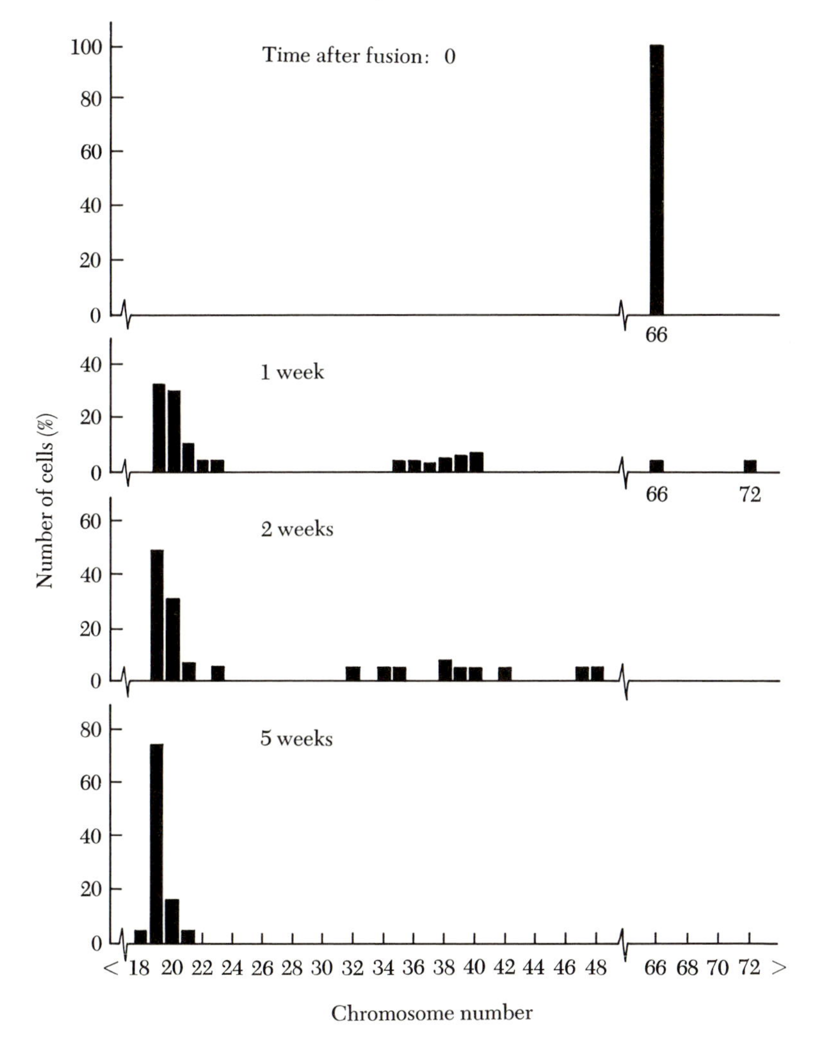

Figure 4-3 Distribution of chromosome numbers in cultures of human-Chinese hamster cell hybrids after various periods of incubation following fusion of the cells. After only one week, the chromosome number is seen to approximate that of the Chinese hamster cell, the majority of the human chromosomes having been lost.

17 seem to be chromosomes of the parental CHO subclone. The two submetacentric chromosomes present, designated H_1 and H_2, are presumably of human origin. Measurements make it seem likely that the larger of these is chromosome 2 and the smaller is either 6 or X. One can, therefore, conclude not only that the inositol and proline genes are unlinked, but that the inositol gene is probably located on human chromosome 2, 6, or X.

These findings illustrate the power of the methodologies that are developing in this rapidly expanding field. One of the critical needs is to obtain large numbers of genetic markers for such studies. Not only can linkage determinations be made but it should be possible to characterize particular chromosomes by the presence or absence of specific genes.

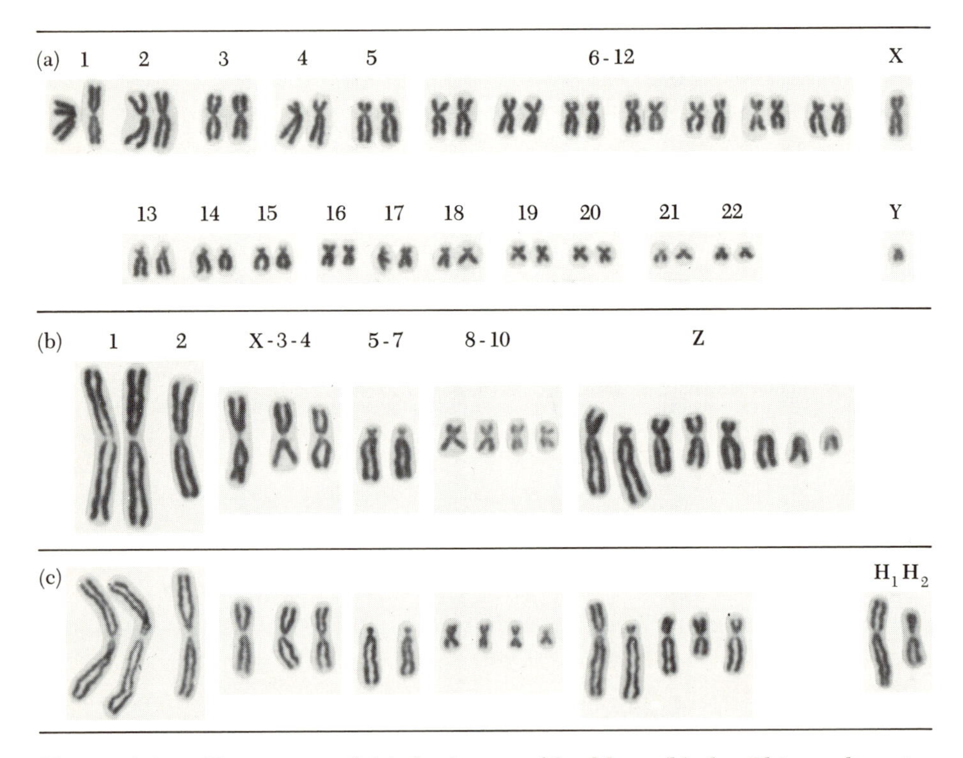

Figure 4-4 Karyotypes of (a) the human fibroblast, (b) the Chinese hamster ovary cell, and (c) the hybrid cell formed by fusion between them. It can be seen that, while the hybrid has lost several of the Chinese hamster chromosomes of group Z, it has retained two (H_1 and H_2) from the human cell parent. Thus, any human genes that can be shown to exist in the hybrid, presumably are located on one of these two chromosomes.

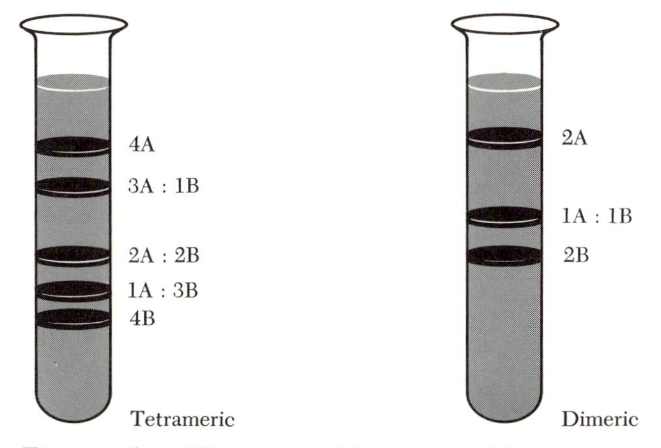

Figure 4-5 *Illustration of the variety of forms an isozyme can assume depending on the total number of subunits in the active enzyme. Thus the two subunits, A and B, can be arranged five ways in the tetramer and three ways in the dimeric enzyme. The relative amounts of each form can vary from tissue to tissue and from species to species. Separation can often be achieved by gel electrophoresis as depicted in the figure.*

A tool which promises to do much in extending the scope of these studies is the analysis by gel electrophoresis of isozymes. Isozymes are enzymes which exist in a variety of forms. The active enzyme consists of an aggregate of similar but not necessarily identical monomeric protein subunits. Although the total number of monomers in any given isozyme is fixed, the ratio of the different kinds of subunits can vary. Thus, a tetrameric isozyme made of two different kinds of subunits can exist in five different forms, depending on the proportion of the two different subunits involved, while one which is a dimer will display three different enzymatically active forms. These different forms are usually separable by electrophoresis and form discrete bands in an electrophoretic gel. (Figure 4-5). The bands can be made visible by treatment with the specific enzyme substrate under conditions such that the product of the enzymatic conversion produces a color change. More than twenty different enzymes can be identified in this way.

Figure 4-6 shows the results of gel electrophoretic determinations for the lactic and malic dehydrogenase isozymes in the two parental cells and in the ino^{+}pro^{-} hybrid containing two presumed human chromosomes, as shown in Figure 4-4. As indicated, the isozyme patterns of the Chinese hamster cell and the human cell are distinct.

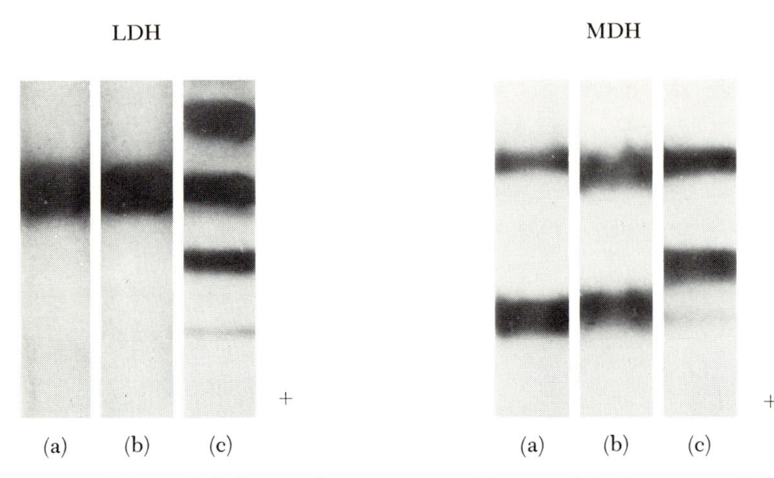

Figure 4-6 Gel electrophoretograms of lactate dehydrogenase (LDH) and malate dehydrogenase (MDH) enzymes obtained from extracts of (a) the Chinese hamster ovary cell, (b) the ino$^+$ pro$^-$ hybrid, and (c) the normal human fibroblast. In both cases, the isozyme pattern of the hybrid is indistinguishable from that of the Chinese hamster cell.

However, the patterns of the hybrids are indistinguishable from that of the CHO cell. The simplest interpretation of this experiment is that, in the human genome, these two enzymes are not linked to the inositol gene and are not carried on the chromosomes designated H_1 and H_2 in Figure 4-4.

Human cells contain a species-specific antigen which causes at least 98% killing of the cells when they are placed in a 1% solution of antiserum from a rabbit previously immunized with these cells. Chinese hamster cells are not affected by this antiserum solution. Consequently, hybrid human-Chinese hamster cells, which have retained a human chromosome bearing this lethal antigen, called A_L, retain their susceptibility to the standard antiserum solution, while those hybrids which have lost this chromosome are immune to such lethal action. It then becomes possible to determine which enzyme activities always accompany A_L activity, and hence are genetically linked to it. Table 4-1 presents data from a series of 76 clones. It demonstrates that the cells with A_L antigen always exhibit activity of the lactic dehydrogenase A isozyme while cells lacking A_L also lack this isozyme. Hence the genes underlying these activities must be linked. The lactic dehydrogenase B isozyme however is unlinked to

A_L. By means of such markers, rapid progress should be forthcoming in determining the linkage groups of the human genes.

One can reasonably expect that cytological identification of the corresponding human chromosomes will then permit each linkage group to be identified with a particular chromosome.

Table 4-1 Tests for Linkage Between A_L and the LDH Isozymes A and B

Cell Phenotype	A_L: / LDH-A:					A_L: / LDH-B:				
	A_L:	+	+	−	−	A_L:	+	+	−	−
	LDH-A:	+	−	+	−	LDH-B:	+	−	+	−
Number of clones found		41	0	0	35		24	17	9	26

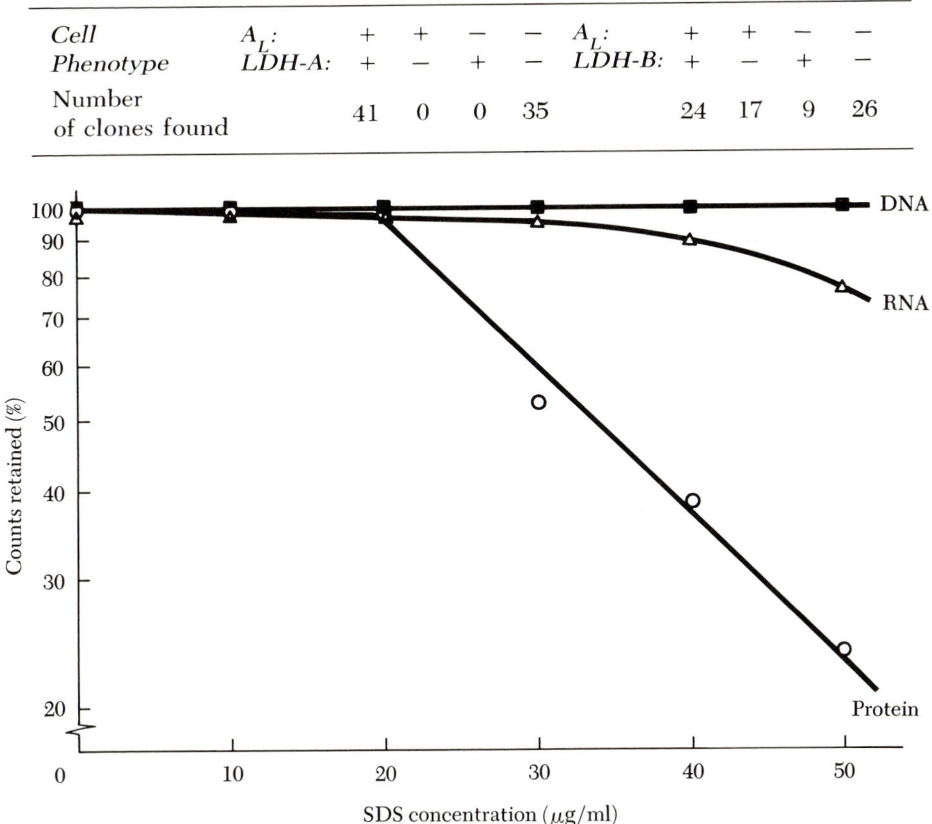

Figure 4-7 Patterns of leakage of radioactively labeled DNA, RNA, and protein from a Chinese hamster cell treated with the detergent sodium dodecyl sulfate (SDS). It can be seen that, while approximately 80% of the cellular protein has leaked out, virtually all of the DNA is retained, indicating that the principal leakage comes from the cytoplasm. If effects such as these can be produced with a variety of cells, and if the resulting cytoplasm-depleted cells can be hybridized with other cells, a useful tool will be available for analysis of cytoplasmic determinants of cell behavior.

4-2 OTHER METHODS WHICH OFFER PROMISE

In addition to these developments, other means of locating genes on chromosomes appear to be close to practicality. One of these is the identification of extra or missing genes in cells with partial or complete monosomic or polysomic chromosomes. This method applies only to genes which obey dosage effect, that is, in which a specific biochemical action is directly proportional to the multiplicity of the particular gene on the chromosome. It would theoretically be possible to identify by biochemical quantitation the extra genes present in polysomic cells, such as those with trisomy 21, as well as the missing genes in cells with a partially or completely deleted chromosome (Figure 4-8). A number of human cell cultures are available with partial or complete monosomies or trisomies derived from patients with particular disease conditions such as Down's syndrome. Conceivably they may also be obtained by cell manipulation in the laboratory. If rapid and quantitative methods could be developed for measuring enzyme activities of a wide variety of biochemical reactions and if gene multiplicity could be determined from such data,

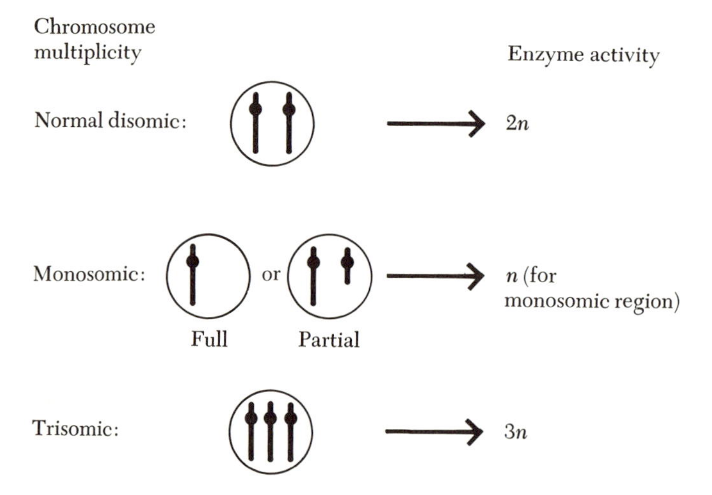

Figure 4-8 Determination of chromosomal multiplicity by quantitation of specific enzyme activities. An enzyme produced from a monosomic locus should be present in an amount approximately half that of the normal disomic state. As the multiplicity of the gene increases, the enzyme activity should increase proportionately. This technique is applicable only for those genes that obey gene-dosage effect.

means for identifying extra and missing genes in such aneuploid cells would be available.

In its simplest form, the problem of gene multiplicity in mammalian cells involves determination of the heterozygous or hemizygous state for recessive genes. In cases where a defective gene can cause human disease, this becomes important practically, as well as theoretically, since heterozygous carriers of a recessive defect may never realize they carry the defective gene until a child with the full-blown disease is born. Often the presence of the heterozygous condition can be demonstrated by a particular biochemical test carried out on the cells of selected tissues. Appendix IV lists some human genetic diseases due to a recessive gene defect, in which the biochemical block has been clearly delineated.

One experimental approach to this problem involves an attempt to monitor, by a relatively simple procedure, the activity of a series of enzymes in a single nonbranching metabolic chain. Thus, if one could add a radioactively labeled precursor to a cell suspension and then determine the amount of radioactivity in each reaction product, the reaction velocity constants for the enzymes in each step of the chain might be calculable. If the procedure were sufficiently rapid and simple, the technique might be applicable to a variety of different metabolic chains and might even be used in the screening of large human and animal populations.

An attempt to realize this possibility was carried out, utilizing the reactions in the metabolism of galactose in human red cells. The reaction chain involved is as follows:

$$A \xrightarrow[K_1]{\text{enzyme 1}} B \xrightarrow[K_2]{\text{enzyme 2}} C \xrightarrow[K_3]{\text{enzyme 3}} D$$

where A, B, C, and D represent, respectively, galactose, galactose-1-phosphate, uridine diphosphogalactose, and uridine diphosphoglucose; enzymes 1, 2, and 3 are galactose kinase, transferase, and epimerase, respectively; and K_1, K_2, and K_3 are the reaction velocity constants for the three enzymes, respectively. Radioactive galactose was added to packed red cells. After a standard incubation period, the test tube was plunged into boiling water to stop the enzymatic reactions, and the contents were chromatographed. The radioactivity corresponding to each of the products indicated in the reaction above was automatically recorded in an instrument which scanned the chromatogram. The results were then fed automatically into a computer which calculated the reaction constants so obtained. Typical results are displayed in Table 4-2.

These results permitted the following conclusions about these reactions in the red cells of human subjects: (1) The values for the reaction velocity constants of the first three steps of the galactose metabolic chain are satisfactorily reproducible in normal human subjects; (2) subjects suffering from the disease galactosemia, which involves a recessive defective gene, have normal values for the galactose kinase but, as expected, have no transferase enzyme activity whatever; (3) persons heterozygous for the galactosemic condition have normal activities for two of the enzymes, but their transferase activity is approximately half that of normal subjects, an indication that the enzyme activity associated with a single normal gene is half of that found in normal homozygous individuals. Thus, this gene displays normal dosage effects. It is of interest in this connection that in normal cells, the rate of removal of the intermediate product galactose-1-phosphate is much higher than its rate of formation. This situation ensures that normally there will never be an accumulation of the galactose-1-phosphate intermediate. There is evidence that the brain damage which occurs in the disease galactosemia is due to the galactose-1-phosphate which accumulates in the absence of transferase enzyme. Thus, in the normal situation the cell protects itself from this toxic action, but the presence of the mutation in the transferase gene breaks down this built-in biochemical protection.

Another simple experimental method is available for determining multiplicity of genes which are required for cell growth and which obey dosage effect. If identical inocula of single cells are added to a series of petri dishes in a minimal growth medium, varying amounts of a competitive metabolic antagonist can be added to the different

Table 4-2 Reaction Velocity Constants (in units of min^{-1}) for the First Three Steps of the Galactose Metabolic Chain in Normal Subjects and Persons with the Heterozygous and Homozygous Galactosemic Defect

	K_1 (Kinase)	K_2 (Transferase)	K_3 (Epimerase)
Normal subjects	0.064	1.18	0.36
Galactosemia heterozygote	0.064	0.46	0.34
Galactosemic patient	0.049	0	Not determined

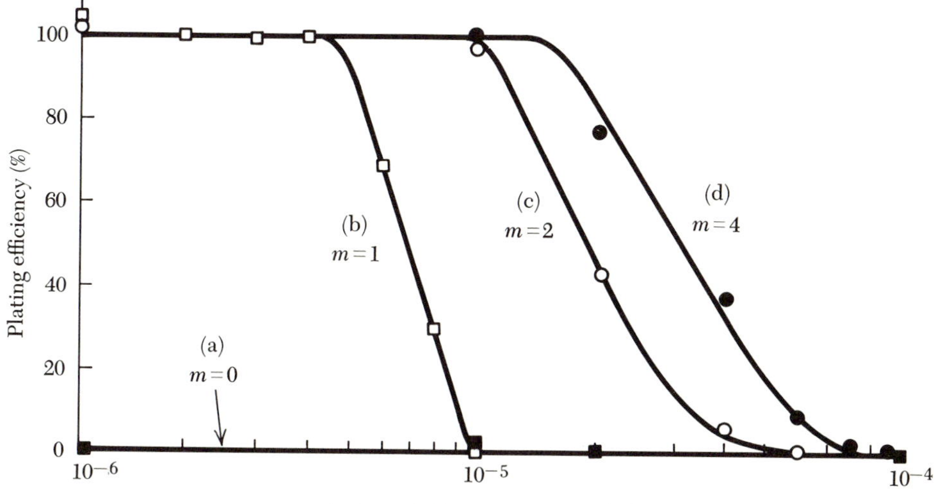

Figure 4-9 Titration of the effect of AZCA on the growth of four Chinese hamster clones differing in the presumed multiplicity (m) of the proline gene, for (a) pro⁻ mutant (m=0), (b) spontaneous revertant of (a) (m=1), (c) normal Chinese hamster cell (m=2), (d) approximately tetraploid Chinese hamster cell (m=4). The fact that the different cell types exhibit such a predictable and differential response to the agent, would appear to make this a valuable technique in determination of gene multiplicity.

plates, and the point at which competition with the normal cellular metabolite becomes great enough to inhibit growth can be determined. In this way, a titration curve can be established for the given cell type. Thus, a cell which has only a single gene for synthesis of the given molecule should be able to withstand less of the antagonist than one which has two genes, and cells with an even higher multiplicity of genes should be even more resistant to the drug. A series of such titration curves revealing the expected differences in resistance is shown in Figure 4-9. In this case, Chinese hamster cells are treated with L-azetidine-2-carboxylic acid (AZCA), which competes with proline and is incorporated instead of proline into proteins, which are thus functionally defective. The data show clearly that, with increased multiplicity of the genes responsible for proline synthesis, the cells become more resistant to the action of the proline antagonist. This titration method produces highly quantitative results and may find further application for titration of multiplicity of genes needed for reproduction. As is discussed in the following paragraphs, methods such as these are also important for identification of heterozygotic carriers of recessive genetic diseases in human populations in order to avoid giving birth to children with serious genetic metabolic defects.

4-3 DNA HYBRIDIZATION AND REITERATION

Multiple genes in *E. coli* are known to exist for a limited number of functions, as in the presence of five- to tenfold repetition of ribosomal genes and existence of duplicated genes for transfer RNA. Mammalian DNA also displays examples of DNA sequence repetition. In some cases identical genes may be reiterated 10 to 100 times, or genes may exist with large amounts of homology, as in the different hemoglobin genes. However, in addition, mammalian DNA exhibits a kind of reiteration different from anything found in *E. coli.*

The process of DNA-DNA and DNA-RNA hybridization[1] can be used in many situations to determine whether appreciable homology exists between a test sequence of nucleic acid, and a given fraction of the genome. Highly repeated DNA sequences will form double-stranded structures very rapidly, and, from the rate of this reaction, one can calculate the degree of reiteration of these sequences. Thus, hybridization experiments have been performed in which short lengths of single-stranded DNA are produced and tested for their ability to re-anneal with each other. It is found that approximately 40% of calf DNA behaves as though it contained sequences which are repeated approximately 10^4 times.

If a highly reiterated DNA sequence were to possess a density of DNA different from that of the rest of the genome as a whole, it should be possible to separate this particular material from the rest of the DNA by means of density-gradient centrifugation, since the highly repeated DNA structure would move to a density position recognizably different from that of the rest of the genome. This is actually observed when mouse DNA is studied in the ultracentrifuge. Approxi-

[1]Hybridization, as applied to nucleic acids, means the bringing together of denatured (i.e., single-stranded) DNAs, or a single-stranded DNA and an RNA under conditions where complementary strands can unite. The resulting double-stranded form can be separated from the remaining single strands by its ability to be retained selectively on a filter or by other physico-chemical procedures.

mately 10% of the total DNA sediments at a position indicating that it has a lighter density than that of the rest of the genome. Moreover, this so-called "satellite DNA," is found to have an enormously high rate of hybridization, and the repeating unit, which appears to consist of 200 nucleotide pairs, is reiterated over a million times. This 200-nucleotide sequence itself consists predominantly of a subunit of six nucleotide pairs, which is repeated many times, and a small proportion of other nucleotides, which are repeated less precisely. The remaining 90% of the mouse DNA seems to be divisible into two categories: 20% of the DNA displays an apparent gene multiplicity of 10^3 to 10^5, and the remaining 70% has apparent gene multiplicities in the neighborhood of 1 to 10.

An experiment carried out by Gall and his coworkers has thrown some light on the function of the highly reiterated satellite DNA of the mouse. Mouse cells were grown in culture in the presence of tritiated thymidine so that a radioactively labeled DNA could then be isolated from these cells. This material was subjected to fractionation by gradient centrifugation. The satellite material was isolated, chopped up into small pieces by shearing stress, and then heat denatured so that short, single-stranded, highly labeled fragments of the satellite DNA resulted. For this part of the experiment, one could also use labeled RNA which is prepared *in vitro* from the satellite DNA by incubation with RNA polymerase. In either case, single stranded, labeled nucleic acid fragments of the mouse satellite DNA are produced.

The next step of the experiment was a bold one. Chromosome preparations of mouse somatic cells were prepared on a microscope slide as in Figure 2-1. After fixation and staining, the DNA in the chromosome was denatured by heat or alkali. It was found that the hard, sharp outline of the mitotic chromosomes became slightly diffuse in appearance, and the stained chromosomes appeared to have increased in thickness. The morphology and integrity of the chromosomes, however, were still well preserved and their ease of recognition was not significantly impaired. Now when the labeled single-stranded DNA or RNA is applied to such a chromosomal preparation, the labeled material hybridizes with its specific complementary regions in the chromosomes. The resulting slide is used for the preparation of an autoradiograph, and the characteristic black marks in the emulsion indicate the point of attachment of the labeled single-stranded material.

When labeled nucleic acid, corresponding to the satellite DNA of the mouse chromosomes, is exposed to the complete karyotype, the material, astonishingly enough, attaches to a particular region of all the chromosomes, namely the centromeric regions (Figure 4-10). Apparently then, this portion of each chromosome contains the DNA which is in a highly reiterated form and presumably does not code for protein. It is conceivable that this material functions in connection with the attachment of the microtubular fibers of the spindle to the chromosomes in the centromeric region, or in the binding of the two chromatids to each other.

4-4 HUMAN IMPLICATIONS

The developments discussed in this chapter imply a variety of direct and indirect human applications. For example, it is conceivable that the methods described for detecting heterozygotes in galactosemia could be extended to include, in the same sample, tests for five or ten of the most common human genetic diseases known to derive from single recessive defects. (This might involve use of other cells, such as lymphocytes, rather than red cells, since the former can be stimulated by phytohemagglutinin to enter into a reproductive cycle and would therefore cause the induction of a much larger variety of enzymes than is present in the red cell.) For each subject so monitored, a series of tubes would be used, each of which receives a different radioactive substrate. Conceivably the use of automated methods and computer-assisted evaluation of the results might make this procedure feasible for routine use for persons considering marriage and child-bearing. Since the disease galactosemia alone occurs in this country in at least 2000 newborns each year, it is evident that a screening procedure of this kind might easily prevent the birth of large numbers of defective children. By furnishing the information to prospective parents about the presence of an appreciable risk of this kind, one enables the parents to make a responsible decision in this vital area. In addition, amniocentesis, carried out during pregnancy, may permit prevention of birth of defective children conceived by two heterozygous carriers. It has already been possible to detect the presence of the gene responsible for the Lesch-Nyhan syndrome on amniotic cells tested in this way. There is little doubt that the number of diseases so detectable, and therefore preventable, will increase.

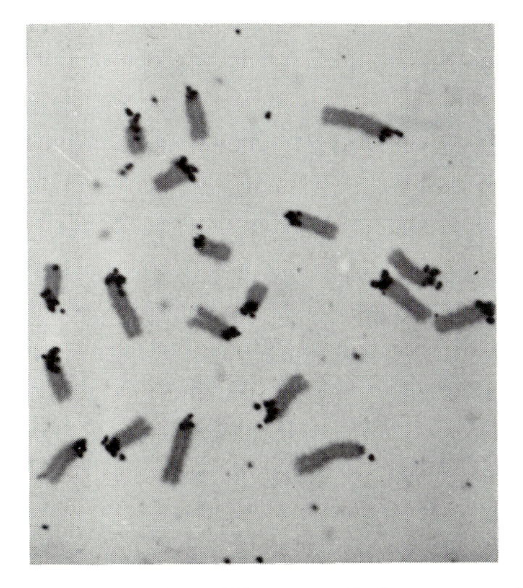

Figure 4-10 *Photograph demonstrating the location in mouse chromosomes of the highly reiterated mouse satellite DNA. Short, single-stranded, highly labeled fragments of mouse satellite DNA were prepared and then hybridized with the DNA of mitotic chromosomes that had been denatured according to the Gall technique. The radioactive material hybridizes specifically with DNA sequences at the centromeric regions of all the chromosomes, thus implicating the satellite DNA in a general centromeric function. (Photograph courtesy of J. Gall.)*

Another application of these techniques stems from the fact that only a tiny fraction of the enormous collection of drugs in the medical *Pharmacopoeia* has a mode of action which is understood biochemically. With the more extensive mapping of metabolic relationships, which should result from the kinds of studies indicated here, the delineation of the specific action of many drugs and pharmacological agents should become facilitated. It may be possible to resolve some of the still-obscure aspects of drug action, relating to the differential sensitivity of particular persons to a drug which is well tolerated by the general population. For example, chloramphenicol, an antibacterial drug, which is highly successful in treatment of many bacterial infections, occasionally produces extremely toxic and often fatal action in particular patients. It may well be that such idiosyncracies are due to the presence of recessive genes in the heterozygous state which betray their presence only when the patient is confronted with the challenge of a drug acting at a specific metabolic site, in much the same way that AZCA revealed heterozygosity in the gene responsible for proline synthesis.

One of the most critical problems that would appear to be soon resolvable is the acquisition of understanding about gene and chromosomal regulation in mammalian cells and tissues. These regulatory processes are raised in mammalian cells to a degree of complexity far surpassing the simple mechanisms, such as those controlling the *lac operon* in *E. coli.*

4-5 SUMMARY

New approaches to the problem of gene mapping in mammalian cells promise to open many possibilities for genetic-biochemical analysis. Cells of different species can now be fused, the hybrids grown up in selective media, and chromosomes unnecessary for growth of the hybrid cell lost. The availability of an animal cell with stable single-gene mutations makes possible correlation of these specific genes with particular human chromosomes. When several markers are available in the same cell, one may also determine linkage between the genes. The problem of gene linkage may also be approached by analysis of gel-electrophoretic patterns of isozymes in the hybrid cells as compared with those of the parental lines.

New methods now allow study of gene multiplicity in the heterozygous, homozygous, and polyploid conditions. For a series of cells with known and different gene multiplicities, one can determine whether the given gene obeys dosage laws. Where it does not, the operation of a genetic regulatory mechanism has been uncovered. Where dosage effect is obeyed, these procedures can be used to determine multiplicity of that gene in a variety of cells.

Experiments utilizing DNA-DNA hybridization have shown mammalian cells to contain sequences of DNA which are repeated as often as 10^6 or more times. DNA hybridization with preparations of mitotic chromosomes has shown one such common sequence to be located at the centromeric region of all of the chromosomes in the mouse.

These studies have application to many human problems. For example, it may be possible to test routinely for the presence of heterozygous defects in couples planning marriage or in the developing human embryo. More extensive mapping of metabolic relationships should facilitate delineation of the specific action of drugs and other pharmacological agents. Finally, these studies should aid in illumination of gene and chromosomal regulation in mammalian cells and tissues.

REFERENCES

Selected papers

Britten, R. J., and D. E. Kohne. Repeated sequences in DNA, *Science* **161**, 529 (1968).

Hill, H. Z., and T. T. Puck. Enzyme kinetics in mammalian cells. II. Simultaneous determination of rate constants for the first three steps of galactose metabolism in red cells, *J. Cell Physiol.* **75**, 49 (1970).

Jones, K. W. Chromosomal and nuclear localization of mouse satellite DNA in individual cells, *Nature* **225**, 912 (1970).

Kao, F. T., and T. T. Puck. Genetics of somatic mammalian cells: Linkage studies with human-Chinese hamster cell hybrids, *Nature* **228**, 329 (1970).

Littlefield, J. W. The use of drug resistant markers to study the hybridization of mouse fibroblasts, *Exptl. Cell Res.* **41**, 190 (1966).

Matsuya, Y., and H. Green. Somatic cell hybrid between the established human line D98 (presumptive HeLa) and 3T3, *Science* **163**, 697 (1969).

Migeon, B. R., and C. S. Miller. Human-mouse somatic cell hybrids with single human chromosome (group E): Link with thymidine kinase activity, *Science* **162**, 1005 (1968).

Nabholz, M., V. Miggiano, and W. Bodmer. Genetic analysis with human-mouse somatic cell hybrids, *Nature* **223**, 358 (1969).

Pardue, M. L., and J. G. Gall. Molecular hybridization of radiation DNA to the DNA of cytological preparations, *Proc. Natl. Acad. Sci.* **64**, 600 (1969).

Pardue, M. L., and J. G. Gall. Chromosome localization of mouse satellite DNA, *Science* **168**, 1356 (1970).

Puck, T. T., and F. T. Kao. Genetics of somatic mammalian cells. VI. Use of an antimetabolite in analysis of gene multiplicity. *Proc. Natl. Acad. Sci.* **60**, 561 (1968).

Puck, T. T., P. Wuthier, C. Jones, and F. T. Kao. Lethal Antigens as Genetic Markers for Study of Human Linkage Groups, *Proc. Natl. Acad. Sci.* **68**, 3102 (1971)

Southern, E. M. Base sequence and evolution of guinea-pig γ-satellite DNA, *Nature* **227**, 794 (1970).

Weiss, M. C., and H. Green. Human-mouse hybrid cell lines containing partial complements of human chromosomes and functioning human genes, *Proc. Natl. Acad. Sci.* **58**, 1104 (1967).

CHAPTER 5

Effect of Radiation on Mammalian Cells

We have seen that the mammalian cell is an elaborate complex, dependent for its functioning on the existence of billions of bits of information stored in highly condensed fashion in specific macromolecules. Ionizing radiation, on the other hand, is a highly random mode of energy transfer, which, when absorbed in matter, causes ejection of electrons from their stable positions within the atoms of the target. Regardless of where any ejected electron was originally located in the atom, the vacancy produced by its interaction with the radiation is ultimately transmitted to the outermost shell, which contains the valence electrons. Therefore, the result of interaction of such radiations with matter is the disruption of the normal valence bonds that make a molecule a stable aggregate. Molecular fragments are produced which can recombine with each other or attack other molecules in a more or less random fashion. Thus, each act of energy absorption from the initial beam causes a series of random molecular bond breakages which ultimately ends when a new, reasonably stable set of molecules has been produced.

It is illuminating in this connection to compare the absorption of ionizing radiation by an inorganic system, such as a beaker of water, and by a living organism. In the former case, the water is already at its lowest possible energy state. As a result of irradiation, water molecules are fragmented, and a large variety of highly energetic moieties such as H, OH, HOH^-, and H_2O_2 are produced. These various species interact with each other and with unchanged water molecules in a rapid and complex set of reactions. However, eventually the net result of all the interactions and recombinations is to restore the structure HOH to all of the hydrogen and oxygen atoms present, since this constitutes the most stable form in which atoms of hydrogen and oxygen can exist. In addition, the information content of the HOH molecule is not very high, so there is little tendency for informational loss, occasioned by the splitting of water molecules, to reverse the effect of the energy gain.

A living system, however, is never in the most stable possible molecular arrangement of all of its constituent atoms. Metabolic processes require an enormous array of molecules in energy-rich arrangements. Therefore, when a living system absorbs irradiation and undergoes random bond breakage, it will, in general, not return to its original form but to some new state characterized by the formation of new molecular structures which, in general, will be at a lower energy level. Moreover, since living systems are made up of macromolecules of highly specific structure, their informational content is huge. Hence random bond breakage introduces great improbability that the original structure will spontaneously be restored (Figure 5-1). Since these molecular structures contain the basic information bank which makes possible the vast array of metabolic processes, the introduction of molecular disorganization can alter living systems in profound ways.

The earliest mammalian radiation biology was carried out on whole animals, and it was soon found that a bewildering array of pathologies could be introduced into mammals by exposure to various doses and kinds of ionizing radiation. Thus, among the consequences of such an experience were leukopenia, gastrointestinal disturbance, capillary fragility and internal bleeding, loss of the ability to form antibodies, cancer formation, aging, and death. It was noted that some tissues of the mammalian body exhibit significant damage even after doses as small as 100 rads[1] while others, such as muscle and brain, seem to resist damage after a radiation exposure 100 times greater. Hence, a theory was formulated to the effect that cells of individual tissues contain their own intrinsic susceptibility to radiation damage. Since tissues like bone marrow can be damaged even after small doses, these were presumed to contain cells of extraordinarily great sensitivity, whereas nerve cells were considered to be intrinsically

[1]A rad is a unit of absorption of ionizing radiation equivalent to approximately 3×10^{-5} calories per gram of tissue.

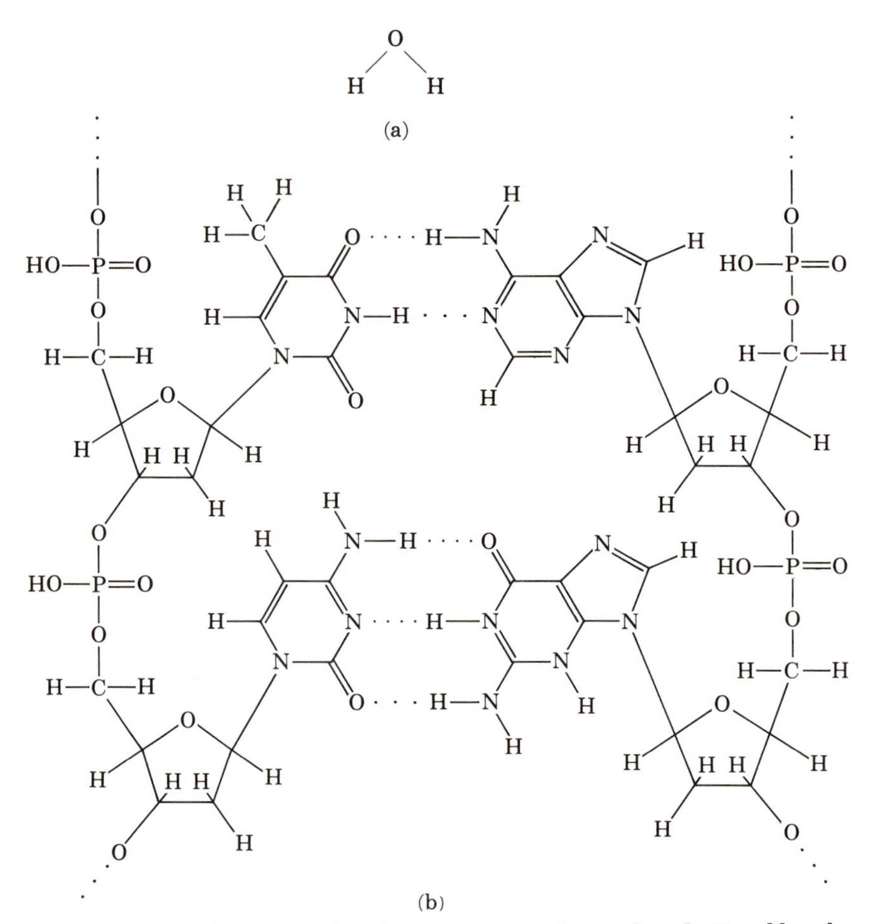

Figure 5-1 *(a) Water molecule, containing only two bonds. Bond breakage can occur only in either of these two positions. If it occurs, energy considerations demand that eventually the same bond structure be restored since all other molecular configurations that can be built of these atoms are much richer in energy than the water molecule. (b) Diagram showing two consecutive structural elements of a gene. Each repeated structural unit contains 78 bonds, any of which can be broken by radiation. This structure is repeated approximately 1000 times in each gene, and even the simplest cells, such as E. coli, contain* 10^4 *such genes. Bond breakage at any point can lead to a new stable molecule, since an enormous number of different rearranged molecules can result from such random breakage, whose energy contents are similar or even less than that of the original molecule. Each of these alterations would represent a change in the biological properties of the cell. Finally, since the genes replicate themselves, such changes would either destroy the replication mechanism or cause an error to be handed down to each of the offspring, thus compounding even more the results of the original informational alteration.*

much more resistant. Many different explanations were invoked to account for this presumed differential radiosensitivity. Among these were hypothetical roles of cyanide-sensitive enzyme systems, sulfur-rich proteins, cellular membranes, toxic chemical substances produced in extracellular body fluids, an increased radiosensitivity of cells in mitosis, and the possible influence of polyploid cells in altering the radiosensitivity of particular tissues. However, no definitive test of the role of any of the general schemes or the specific cellular mechanisms proposed was forthcoming, and, at least as recently as 1960, reviewers in examining the situation were unable to reach a clear understanding of the basic underlying factors.

Understanding of the mechanisms of mammalian radiation injury from whole-animal studies alone was inconclusive because of the difficulty in disentangling effects due to primary radiation injury from those resulting secondarily from the altered metabolic patterns produced in the treated tissues, and because the complexities of cellular dynamics in the whole animal tend to obscure the basic processes occurring at the cellular level. With the development of quantitative methods for studies on mammalian cells *in culture*, a new era in mammalian radiation biology became possible, in which actions of radiation at the cellular level could be defined and quantitated. Application of these studies to processes occurring in the whole animal has resulted in simplification in understanding of the mammalian radiation syndrome and has provided means for new studies involving both whole animals and tissue cultures. The results of such studies also have had important practical applications in treating radiation injury, increasing the usefulness of radiation as a therapeutic agent in cancer, and in providing understanding about normal cell development in mammalian tissues.

The principal questions which one would like to answer about the action of radiation on mammalian cells are as follows: (1) What is the basic dose of ionizing radiation which causes cessation of mammalian cell reproduction? (2) Does this dose vary significantly for mammalian cells in different tissue environments? (3) What is the primary target for this action of radiation? (4) What is the quantitative efficiency of ionizing radiation in producing mutagenesis in mammalian cells? (5) What effects can irradiation have on differentiation activities of mammalian cells? (6) What is the efficiency of production of carcinogenesis in various types of mammalian cells exposed to ionizing radiation?

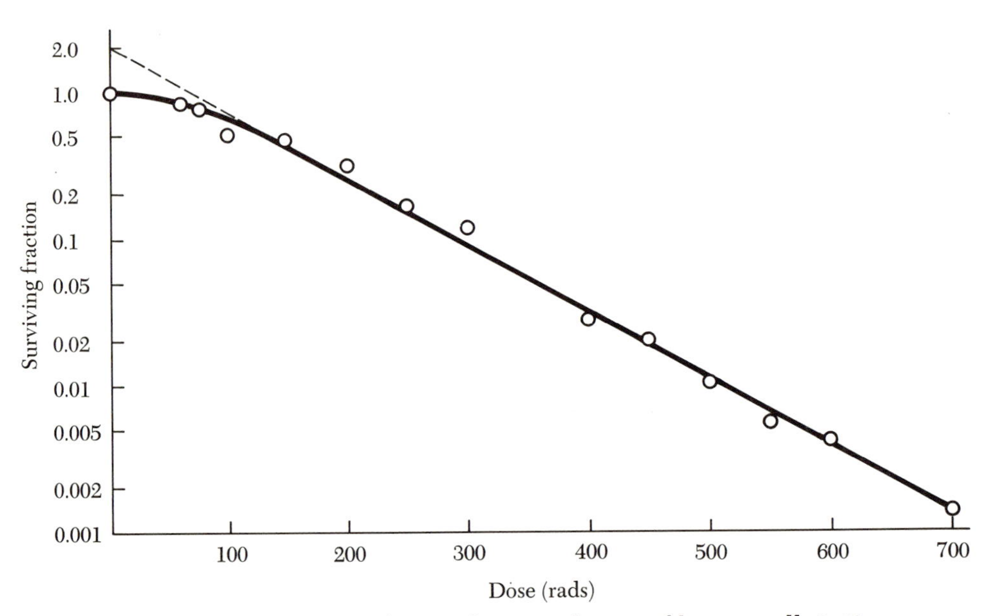

Figure 5-2 *Single-cell survival curve for x irradiation of human cells in tissue culture. All mammalian cells so far tested give curves similar to or only slightly different from this one.*

5-1 SINGLE-CELL SURVIVAL CURVES IN CULTURE

A primary action of radiation in inhibiting cellular reproduction had been proposed several times for mammalian cells. However, many investigators were led to believe that the basic dose needed to inhibit cell reproduction lay in the neighborhood of 50,000 rads. Since the mean lethal dose for killing most mammals is approximately 400 rads, this effect in itself seemed to play a relatively small role in determining the subsequent history of an irradiated animal. However, with the development of means for accurate measurement of survival curves for mammalian cells *in culture*, a precise measurement of the mean dose for reproductive death became possible. The results of such a study are shown in Figure 5-2.

Several noteworthy features are present in this survival curve. First it corresponds to one of the classical survival-curve shapes, namely an initial shoulder followed by a linearly logarithmic decline in the number of survivors as the dose increases. If the straight portion of the curve is extrapolated back until it hits the axis, the intersection occurs close to 2, a form of behavior which would be exhibited if two independent radiation absorption events were required in order for cell death to occur. The most astonishing feature of this curve, however, is the extremely small value of the mean lethal dose, D^0, which is defined as the amount of radiation needed to reduce the number of viable cells to 37% as measured in the linear portion of the curve. (It should be emphasized here that we are using the word "kill" to imply reproductive death, a notation borrowed from microbiology.) The previous estimate of 50,000 rads for D^0 was arrived at on the basis of various experiments including analogy with the dose needed to kill bacteria (*E. coli*) or protozoa (paramecium). In contrast, however, the experiment summarized in Figure 5-2 indicates the mean lethal dose to be only 100 rads. This quantity of radiation absorption represents an amount of energy equivalent to a temperature rise of approximately 0.003 centigrade degrees.

These initial experiments were carried out on the carcinomatous HeLa cells. It was necessary, therefore, to repeat this determination on cells originating in normal tissues, in order to ascertain whether the high sensitivity might simply be a characteristic of certain cancer cells. New biopsies, therefore, were taken from a series of tissues from man and experimental animals. The survival curves obtained closely approximated that shown in Figure 5-2, and indeed some of the normal mammalian cells tested revealed a mean lethal dose of only 75 rads, indicating even greater sensitivity than the HeLa cell. Virtually all mammalian cells tested yielded similar survival curves, almost all of which had an initial shoulder and exhibited D^0 values clustering around 100 rads (Table 5-1). Cells taken from supposedly highly sensitive tissues, such as bone marrow, displayed virtually the same radiosensitivity as those taken from any other part of the body when tested by this *in vitro* method.

5-2 X-RAY SURVIVAL CURVES OF CELLS IN THE INTACT ANIMAL

It was then necessary to test critically whether the survival curves obtained *in vitro* actually mirrored the loss of reproductive capacity undergone by cells irradiated *in vivo*. The curves, like those shown in Figure 5-2, might after all simply reflect some aspect of the unnatural environment in which mammalian cells find themselves in an *in vitro* situation.

Table 5-1 Approximate Values of D^0 for Various Types of Mammalian Cells Measured in Vivo and Vitro

Cells Studied	D^0 *(rads)*
Normal human cells; Aneuploid human cells from normal and malignant tissue; Normal Chinese hamster, or monkey kidney cell, *in vitro*	125 (50-170 range)
Mouse leukemia cell *in vivo*	152
Mouse fertilized ovum *in vivo*	125
Mouse bone marrow *in vivo*	105
Mouse bone marrow *in vitro*	105

A number of different ingenious approaches were devised by investigators in several laboratories to test this point. Hewitt and Wilson in England were first to carry out a definitive experiment. They were engaged in determining how many mouse leukemia cells were required to initiate the disease in normal mice. They found that intravenous injection of two cells taken from a mouse with advanced leukemia was sufficient to transmit the disease to a new animal. It then became a simple matter to compare the x-ray survival curves for such leukemic cells, *in vivo* and *in vitro*. On the one hand, the cells were taken from the leukemic mouse and irradiated *in vitro*; then the number of such irradiated cells needed to transmit the leukemia to a new mouse was determined. In a parallel experiment the leukemic mouse itself was irradiated and again the number of such irradiated leukemic cells needed to introduce a proliferating disease into a normal animal was determined. It was found that the survival curves obtained by each procedure were identical and also agreed very closely with the survival curve for the HeLa cell which had originally been determined by means of the single-cell colony-forming technique.

In a second test of the identity of *in vitro* and *in vivo* survival curves of mammalian cells, a calculation was made for the survival to be expected when newly fertilized mouse ova were irradiated with 500 rads, assuming that the survival curves of Figure 5-2 and that found by Hewitt and Wilson applied to these mouse cells. On this basis, one could compute the fraction of such fertilized ova that should proceed to the formation of a visible embryonic colony after eight days. An experimental determination of this figure had been carried out with great care by Lliane Russell and her collaborators at Oak Ridge. The calculated and experimental values agreed completely, and indicated a reduction of 80% in the capacity of such ova to grow to visible embryonic colonies.

Perhaps the most elegant experiment of all, however, was that which was carried out by McCulloch and Till in Toronto. It had been known that mice subjected to the minimum lethal dose of total-body x irradiation could be saved from death by injection of nucleated bone marrow cells from a normal mouse. The Canadian experimenters looked into the dynamics of this effect and discovered that injection of normal bone marrow cells caused the development in the spleen of discrete colonies of proliferating cells, Figure 5-3. By demonstrating a linear relationship between the number of injected cells and the number of spleen colonies developing, they were able to demonstrate that each such colony originated from a single cell. Hence, the counting of these colonies furnished a simple and direct method for quantitating relative numbers of viable colony-forming cells present in an initial population. They, therefore, again performed an experiment like that of Hewitt and Wilson, in which they first irradiated bone marrow cells *in vitro* before injecting them into one set of lethally irradiated mice and then, in a parallel experiment, irradiated a set of donor mice before removing bone marrow cells for injection into the lethally irradiated test mice. Again they obtained two survival curves

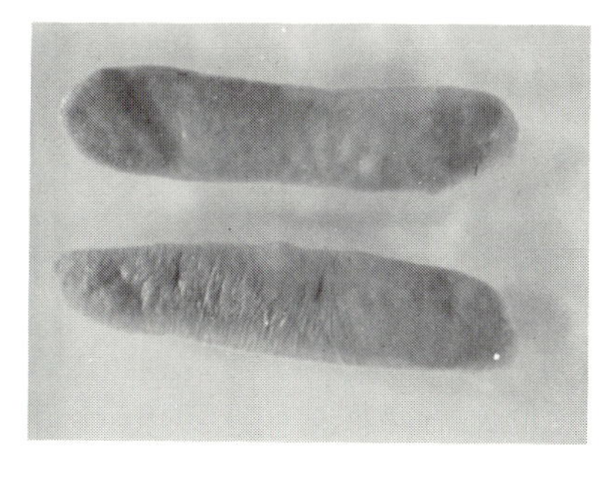

(a)

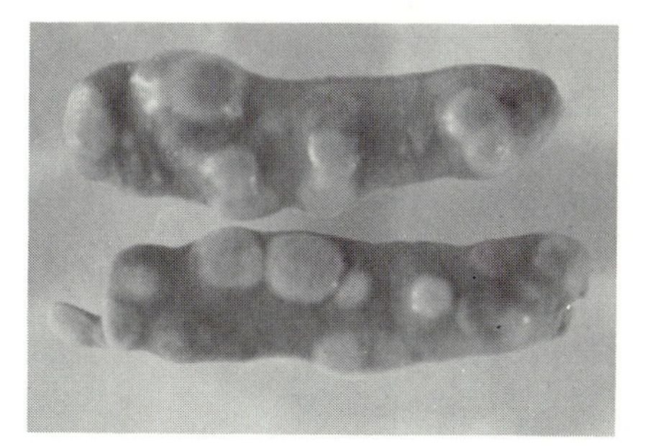

(b)

Figure 5-3 (a) The spleen of a lethally irradiated mouse that has received no injection of bone marrow cells. (b) Spleens of lethally irradiated mice that received injections of bone marrow cells immediately after irradiation. These cells have formed active colonies in the spleen that will reconstitute the depleted bone marrow and thus rescue the mice from radiation death.

which paralleled each other very closely and also closely resembled the survival curve obtained by Hewitt and Wilson and by the original cell plating experiments. On the basis of these experiments, then, it seems reasonable to conclude that there is no essential difference between *in vivo* and *in vitro* radiation survival curves.

5-3 GIANT CELL FORMATION

As we have seen, the absorption of a mean lethal dose of ionizing radiation by a mammalian cell represents an amount of energy so small that it is not obvious how it could kill the cell. It is of considerable interest, therefore, to examine a variety of metabolic reactions in order to determine which of these can continue even after the cell has absorbed an amount of radiation far in excess of the lethal dose. Such studies reveal that even after 1000 rads of x irradiation, mammalian cells continue to carry on a wide variety of biochemical activities, despite the fact that their ability to form colonies has been permanently terminated. Thus, metabolites of all kinds continue to be taken up from nutrient medium, sugars are oxidized to yield energy, and synthesis of protein, DNA, and RNA continues at an active rate in such cells.

One searching experiment that was carried out to test the integrity of biosynthetic systems in a supralethally irradiated mammalian cell involved irradiating a cell population and then treating it with a virus which it had never before encountered. It is well known that viruses require integrity of extensive regions of the cell machinery, including the presence of many cell enzymes, transfer RNA, activated amino acids, ribosomes, and energy-conversion systems in order for the virus to carry out its own invasion and replication. Therefore, successful biosynthesis of new virus in such a radiated cell would testify to the continued activity of large portions of the cell's metabolic apparatus. Such experiments revealed that not only was good virus replication achieved but that often one could obtain approximately 50 times more virus from an irradiated cell than from a normal one. The implication of such an experiment, then, is that the radiation damage is confined to some particular structure(s) of the cell.

If irradiated cells can indeed metabolize without dividing, one might expect to obtain enlarged cells due to continued growth without reproduction. This expectation was indeed realized. Figure 5-4 shows a petri dish in which a number of small dark areas are observable.

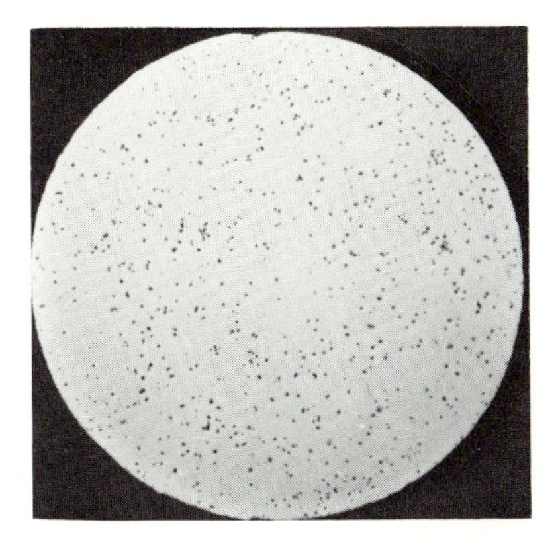

Figure 5-4 *Petri dish containing single cells that have received a dose of x irradiation sufficient to leave no survivors capable of reproduction. Much of the metabolic machinery of the cell, however, is unaffected by this low dose of radiation, and each cell therefore becomes a giant, visible to the unaided eye.*

These might at first be mistaken for small colonies. On the contrary, however, each of these is a single giant cell, the result of cell growth without division. Figure 5-5(a) shows a representative giant cell, and Figure 5-5(b) a typical field in an irradiated dish, in which a colony has developed from a single surviving cell. Around the colony, however, are cells which have been killed by the radiation and which have continued to metabolize and increase in size so as to become giants. These giant cells can achieve a diameter of approximately one millimeter when spread on a glass petri dish and are readily visible to the unaided eye. They form a fascinating biological population which can be used for a variety of different kinds of experiments.

5-4 EVIDENCE THAT THE CHROMOSOMES FORM THE TARGET SITE FOR THE LETHAL ACTION

The question arises as to the identification of the target site within the cell responsible for reproductive death after such small radiation exposures. It must be remembered that when a cell is irradiated, the energy absorption and consequent bond breakage will be randomly distributed throughout all the molecules within the cell. However, the concept of a critical target implies that certain structures exist in which these events cause much more serious damage than do the same

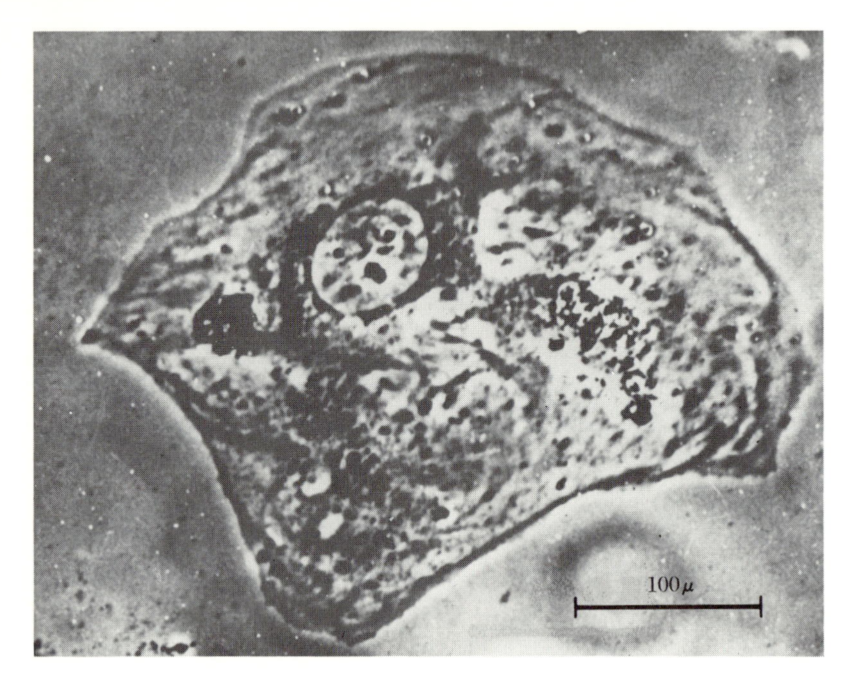

(a)

Figure 5-5 (a) *A single giant cell from a plate similar to the plate in Figure 5-4.* (b) *Photomicrograph of a normal colony that has grown up amongst the irradiated giant cells. The photomicrograph demonstrates that the giant cells achieve diameters 7-10 times that of the normal cell.*

processes in other parts of the cell. The following considerations would appear to demonstrate conclusively that the principal site of the lethal damage lies in the chromosomes and specifically in their DNA.

If the chromosomes are indeed the site at which the events leading to cell lethality principally occur, one might expect to be able to demonstrate changes in chromosomal structure after a radiation experience in the neighborhood of the mean lethal dose. However, processes exist which can cause the broken fragments of chromosomes to reunite. Moreover, since the chromosomes of mammalian cells can be visualized only at mitosis, one cannot be sure that all of the radiation-damaged cells will progress normally to mitosis, so that one might easily miss a certain proportion of the population with damaged chromosomes. Despite the existence of these factors, which operate to decrease the apparent yield of chromosomal anomalies produced by

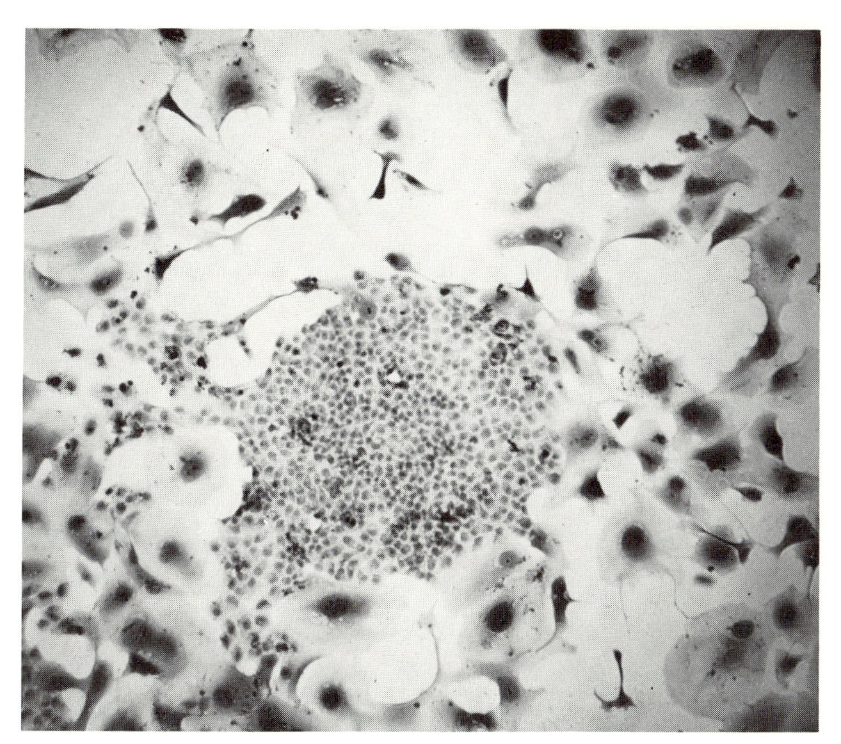

(b)

ionizing radiation, careful scoring of the chromosomal damage reveals that these structures reflect visible changes as a result of irradiation in the mean lethal dose range (Figure 5-6). Visible chromosomal changes can be seen at and below the mean lethal dose but no other cell modification has been seen at such low doses.

Another kind of experiment leads to the same conclusion. Cells were grown in three different vessels with tritiated thymidine present in the first, tritiated uridine in the second, and tritiated leucine in the third. These metabolites are incorporated, respectively, into the DNA, the RNA, and the protein of the cell. Radioactive disintegration taking place in the molecules containing the tritium results in bond breakage by recoil at the point of attachment of the tritium itself and by peripheral ionization from the emanated electron (which is sharply localized because of the extremely weak energy of the tritium radioactivity which is confined within a radius of less than 1 micron). Cell suspensions were grown in various concentrations of each isotopical-

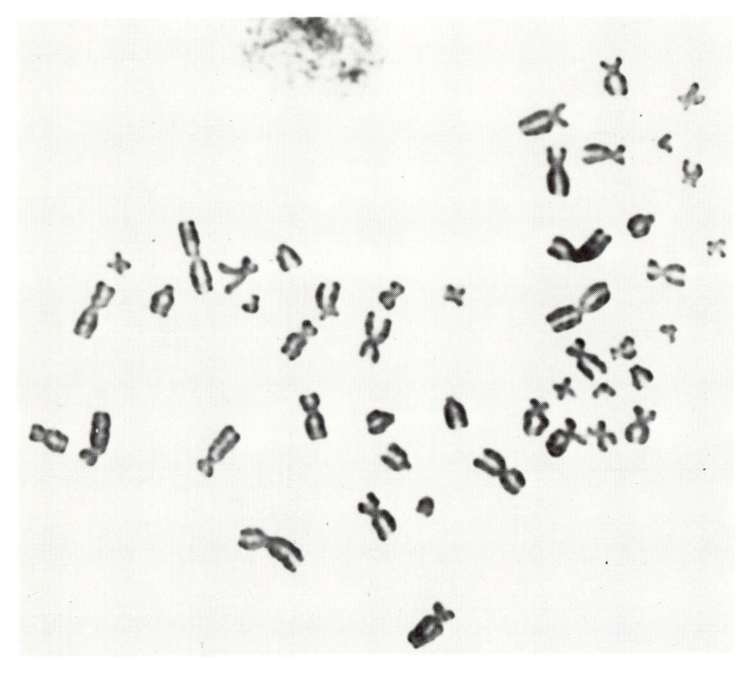

(a)

Figure 5-6 Demonstration of x-irradiation damage to human chromosomes at doses in the range of the mean lethal dose. (a) Normal human chromosomes from an unirradiated cell. (b) Chromosomes from a cell irradiated with 75 rads showing simple breaks and deletions. (c) Chromosomes from a cell irradiated as in (b) above, but in which random rejoining of the broken chromosome ends has produced a variety of abnormal structures. (d) Chromosomes from a cell irradiated with 150 rads, a value slightly above the mean lethal dose. The rejoining of broken chromosomes has produced rings as well as other bizarre configurations.

ly labeled metabolite, and samples were withdrawn after a standard incubation period to determine the number of cell survivors by means of the single-cell plating method. At the same time, other samples were taken for study by radioautography. Cells were placed on microscope slides, all of the small molecular materials remaining were washed away and a photographic emulsion was applied so that radioactivity that had become incorporated into large molecules produced disintegrations in the sensitive photographic emulsion. Grain counts were made on such radioautographs and the number of cell survivors was then plotted against the average number of grain counts per cell to produce the survival curves shown in Figure 5-7.

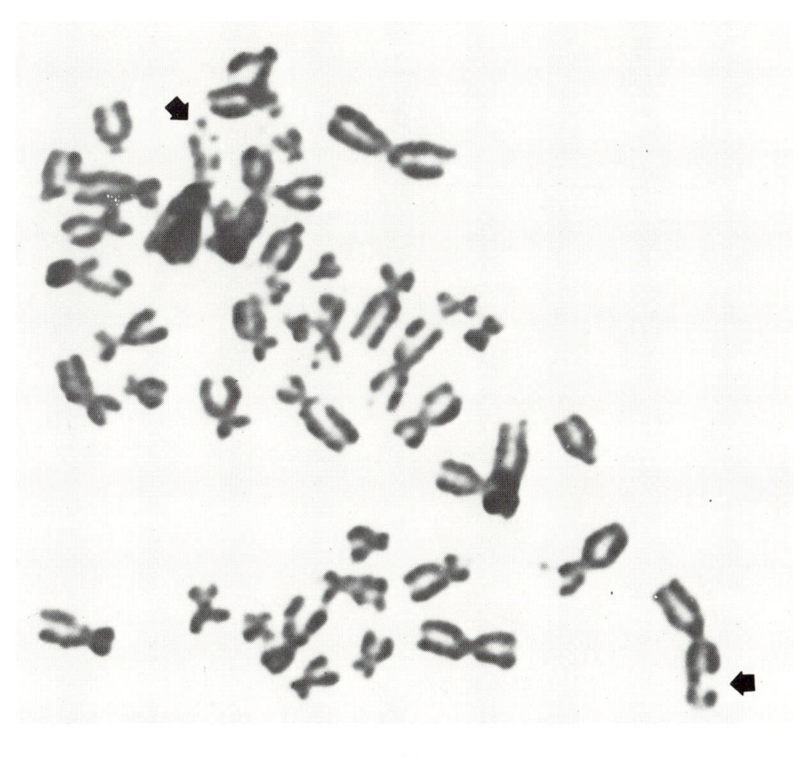

(b)

The data demonstrate, clearly, that for the same total amount of radioisotope incorporation, cell death is maximal when it occurs in DNA, considerably less in RNA, and least of all in protein.

Still another experiment leading to the same conclusion was carried out in an elegant fashion by Szybalski and Opara-Kubinska. These investigators first demonstrated that the incorporation of bromodeoxyuridine (BUdR) into the DNA of mammalian cells increased their sensitivity to killing by x irradiation. A similar effect was noted in bacterial systems. Quantitative comparison of the results in the two systems revealed that the increase in mammalian cell lethality by x irradiation, produced by incorporation of a given amount of BUdR into its DNA, was exactly equal to the increased inactivation of pure bacterial DNA acting as a genetic transforming agent after it had incorporated the same amount of BUdR. This result would be an amazing coincidence unless the DNA is indeed the principal target site for the lethal effect of ionizing radiation on mammalian cells. The

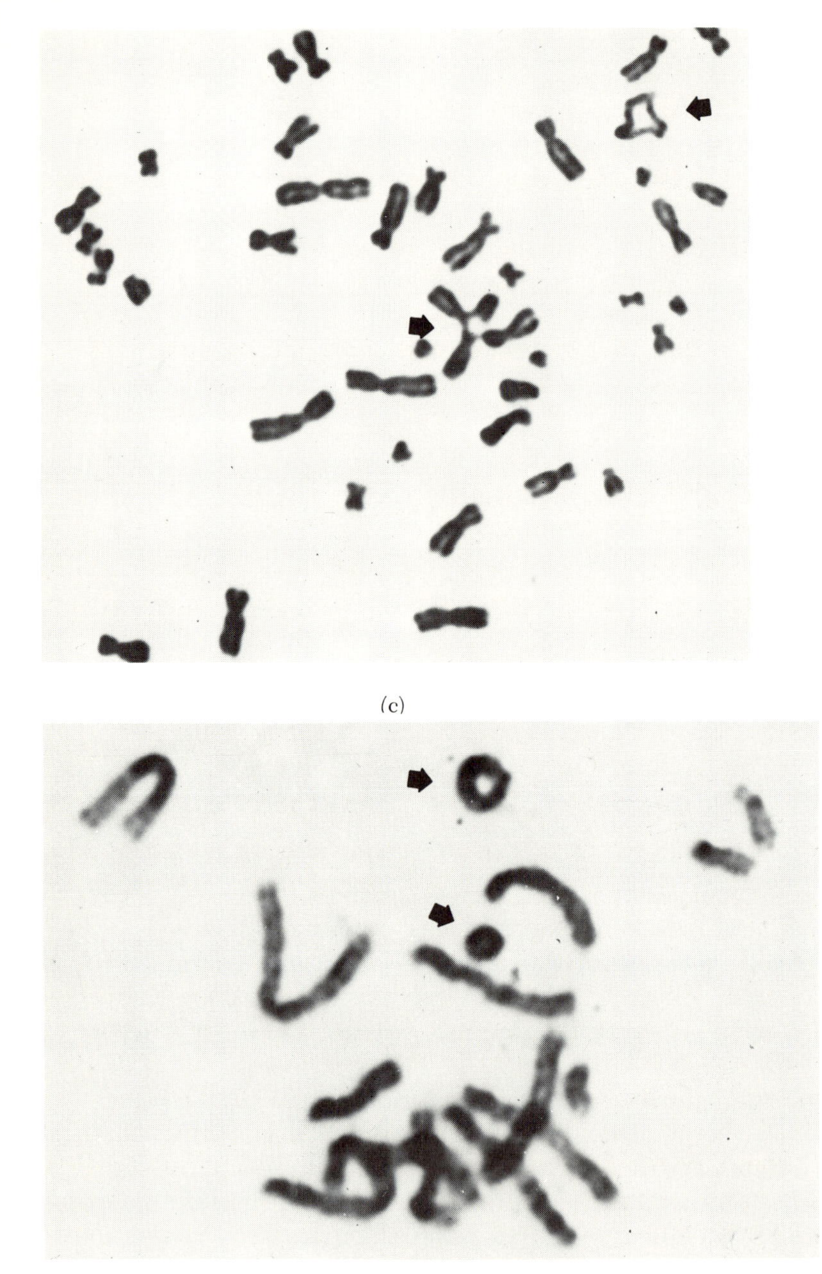

(c)

(d)

Figure 5-6 (continued)

concordance of these three independent types of experiments would appear to prove the thesis beyond doubt.

The question then arises as to the exact mechanism by which cell killing is achieved as a result of radiation-induced damage to DNA. Ionizing radiation is known to produce both single-gene mutations and a variety of chromosomal anomalies in bacteria and organisms such as Drosophila. Mechanisms exist in the cell to repair radiation damage at the level of both the gene and of the chromosome. Repair of single strands of the DNA occurs by excision of damaged regions and resynthesis of the strand in accordance with the sequence of the complementary strand. In addition, broken ends of chromosomes can

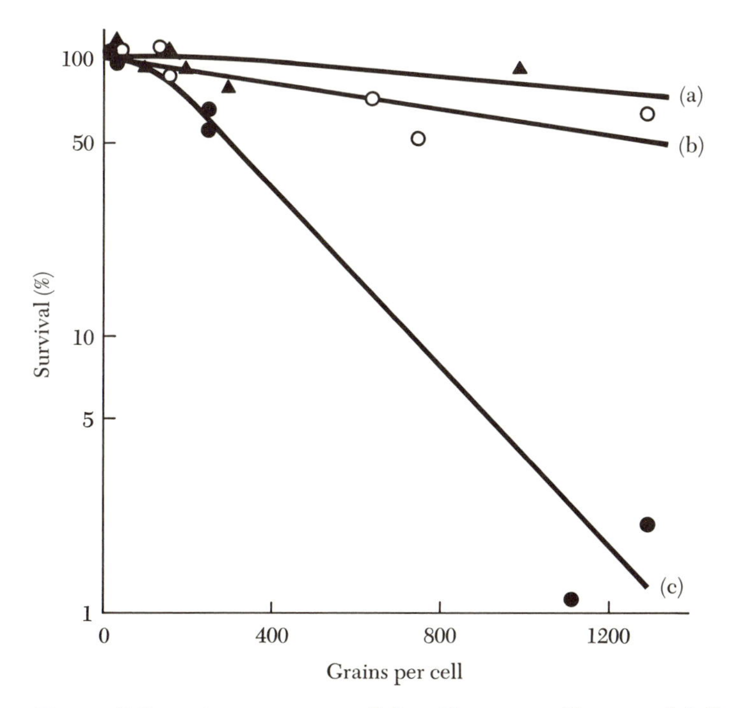

Figure 5-7 A comparison of the effects on cell survival following macromolecular incorporation of H^3 into (a) protein, by use of H^3-leucine; (b) RNA, by H^3-uridine; and (c) DNA, by H^3-thymidine. The abscissa is expressed in terms of average number of grain counts per cell. The curve for H^3-thymidine indicates a multihit process with an extrapolation number of approximately 1.5, which is close to the value of 2 derived from similar survival curves for x irradiation. These data indicate the greater sensitivity of cells to reproductive death when the radioactive decay occurs in DNA molecules as opposed to RNA or protein.

be rejoined to each other. If all the needed steps of each process occur in the right order, a completely normal cell may result. If steps are omitted or executed erroneously, or if nonhomologous chromosome ends are rejoined, gene mutations or chromosomal aberrations or both may occur. The cell killing and mutation production, which are normally observed, represent damage exceeding the capacity of the repair processes.[1]

As explained in Chapter 3, the mutation efficiency for single-gene mutations in mammalian cells was measured and found to be 4×10^{-8} mutations per gene locus per rad. Since mammalian cells are diploid, and most single-gene mutations are recessive, the probability of hitting the corresponding gene in both homologous chromosomes with a dose of only a few hundred rads is negligibly small. Therefore, single-gene mutations can be ignored in attempting to understand the nature of the lethal action of ionizing radiation for these cells. One hypothesis to explain the nature of cell reproductive death following exposure to x irradiation is as follows: The basic primary lesion produced by irradiation is the production of a chromosome break, presumably involving a discontinuity in the DNA. If only one chromosomal break is introduced into a cell, sooner or later the two broken ends will find each other, and, after they have been resealed, the chromosome in question will retain no evidence of its irradiation experience.

When the cell has been exposed to a large enough dose of radiation so that multiple breaks occur, the randomness inherent in the resealing process will result in a certain amount of random rejoining of chromosomal fragments. A variety of different chromosomal and chromatid translocations and deletions will result. Some of the chromosomal configurations so produced may still be viable. However, others may cause cell death because the chromosomal patterns so produced may be incompatible with a state of continuing cell reproduction. This could come about in at least two ways. Mitotic chromosomal bridges can be formed such that homologous centromeres drawn to opposite poles of the cell are joined by the bridges in such a

[1]Certain agents (e.g., caffeine) inhibit repair processes, and, in these cases, more extensive cell damage occurs. Mutation can produce similar behavior. Of particular interest in this connection is the genetic disease in man, xeroderma pigmentosa, in which the skin displays extraordinary sensitivity to ultraviolet irradiation. *In vitro* studies on skin cells have demonstrated that one of the repair mechanisms is absent in these cells, a condition recalling that of the "sensitive" mutants of *E. coli,* which also are highly sensitive to radiation because of loss of a repair enzyme.

way as to prevent separation of the daughter cells. Bridges of this kind, which prevent the consummation of mitosis, are frequently seen in irradiated cells (Figure 5-8). In addition, such bridges might prevent activation of gene sequences either for replication or RNA synthesis that would be required to transport the cell from stage to stage along its life cycle (see Chapter 6).

Some of the evidence in favor of this theory is as follows:

(1) For low doses in the region of the shoulder, the principal chromosomal lesions observed are simple breaks. With larger doses, abnormal restitutions predominate.
(2) Mitotic bridges are seen with high frequency (approximately 10 to 20%) in cells irradiated with four or five mean lethal doses.

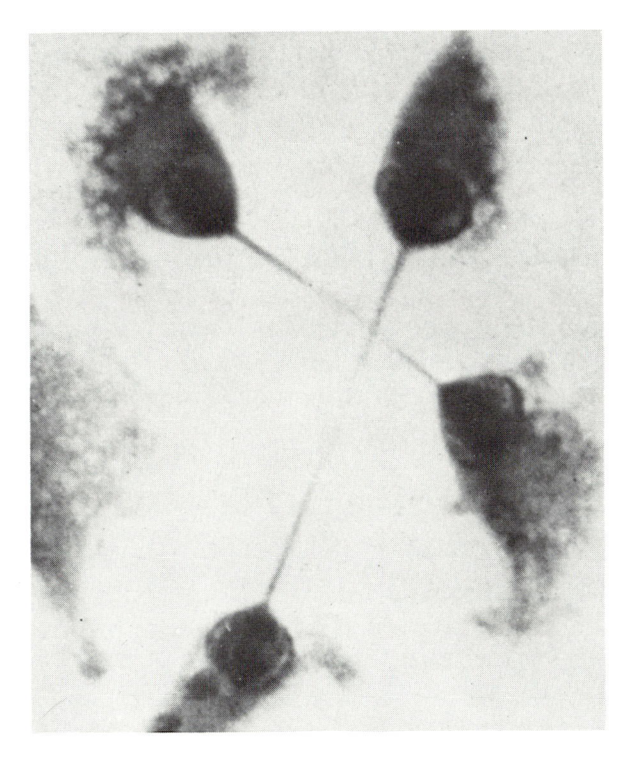

Figure 5-8 Micrograph showing mitotic bridges formed between the separating mitotic nuclei in an irradiated cell population. This type of abnormality, which is observed in many kinds of irradiated cells, prevents the normal separation of the daughter cells and thus the completion of mitosis. (Photograph courtesy of J. F. Whitfield.)

(3) Elkind and his coworkers demonstrated that with small doses of radiation that do not exceed the shoulder region of the survival curve, cells, if given a period of 4 hours of incubation, will completely repair their radiation damage so that if a second radiation dose is now administered, the effect on the cells will be the same as though the first irradiation episode had not occurred. However, if the time between the doses is much shorter, the two doses are additive, behaving as though a single dose equal to the sum of the separate radiation experiences had been administered.

(4) Finally, this hypothesis explains the very high sensitivity of mammalian cells to reproductive death by x irradiation. It postulates that the essential lethal action is a mechanical one involving the chromosomal structure rather than specific gene inactivation. Since almost any part of the genetic structure can take part in these lethal processes, the effective target site includes the whole chromosomal complement. The much larger size of this structure in mammalian cells, as compared to *E. coli,* and the great dependence in the mammalian cell on maintenance of a particular ordering of the various chromosomal elements causes it to have the greatly increased sensitivity as compared to the simpler bacterial cells. In accordance with this expectation, it has been shown that there is indeed a rough parallelism between total DNA content and radiosensitivity in various cell types. The theory here advanced would demand this as a fairly general relationship but could explain occasional departures from it because of differences in the kinetics of the biochemical processes responsible for chromosome restitution.

5-5 APPLICATION TO THE MAMMALIAN RADIATION SYNDROME

Theoretical considerations. The considerations discussed in the foregoing sections of this chapter allow great simplification in understanding the sequence of events attending the exposure of mammals to high energy radiation. While the intricacies of mammalian organization and the randomness of bond breakage resulting from ionizing radiation contribute to making the detailed consequences of irradiation on all or part of the mammalian body highly complex, the major outlines of the process have been considerably clarified, and much that hitherto seemed obscure appears now to be reasonably well understood.

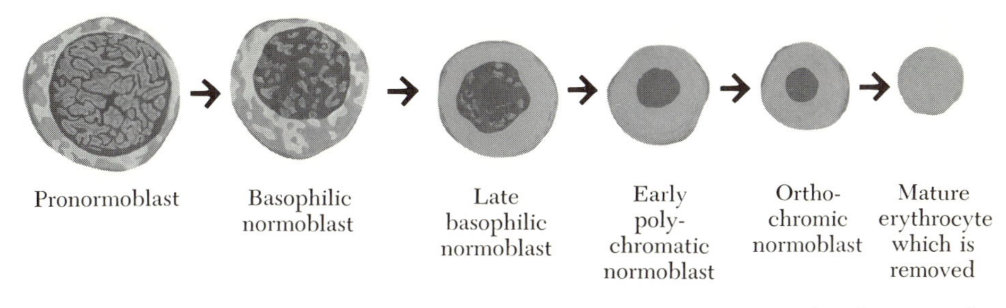

Figure 5-9 *Normal progression of a red blood cell through its developmental cycle. The young cells reproduce and exhibit a large nucleus and high rates of RNA and protein synthesis. As the cells mature, reproduction ceases, the nucleus begins to shrink, and RNA and protein synthesis become diminished. The synthesis of a specific protein, hemoglobin, however, is greatly accelerated. Finally, the nucleus is extruded from the cell. After a period of time, the cell itself is removed from the system.*

As mentioned earlier, the difference in the degree of pathology exhibited by various tissues was originally attributed to intrinsic differences in the susceptibilities to killing of cells of different tissues. It was necessary to postulate differences of 1000-fold or more in the sensitivity of cells of different types. However, despite many studies designed to elucidate the biochemical basis of this enormous range of difference in cell sensitivity, no clear picture emerged that could effectively explain the observations, predict future experiments, or supply an effective guide for the application of radiation to practical problems such as tumor therapy.

With the demonstration of the very low dose of ionizing radiation needed to kill mammalian cells reproductively and the finding that cells from virtually all tissues had very similar sensitivities to reproductive death, it became possible to reexamine the mammalian radiation syndrome in terms of the theory that the primary action of radiation is a general inhibition of cell reproduction rather than a series of different specific actions on the cells in different tissues.

Consider the dynamics of the somatic cells of any tissue in the mammalian body. Cellular structures and biochemical activities are constantly undergoing specific progressions that finally result in an end-product cell that is stable or eventually is removed by the body. An example showing the progressive development of red blood cells, which are formed in the bone marrow, is depicted in Figure 5-9. In the mature adult, the rate of cell destruction is exactly balanced by replenishment through cell reproduction. Therefore, if, as a first approximation, direct transfer of cells between different organs is excluded, we can consider each organ as a compartment in which reproduction and removal are exactly balanced for each cell type. If ionizing radiation then inhibits only cell reproduction and the ordinary processes of cell removal continue unchanged, the number of cells remaining must drop steadily after irradiation. This drop would represent the rate at which the normal cell-removal processes go on when they are no longer compensated by an equal supply of cells through reproduction.

Experimental tests. A critical experimental test of this theory becomes possible when one considers how the number of cells remaining in a given tissue compartment should change with time after the animal has been exposed to various doses of irradiation. Consider the following experiment: A series of mice of the same age, weight, genetic constitution, and sex are given a dose of total-body x irradiation. At varying intervals of time, the femur of the hind leg is removed, its total cell contents quantitatively washed out, and the number of nucleated cells is carefully counted. As a result, one can then obtain for each dose a depletion curve in which a logarithm of the number of cells remaining in the femur is plotted as a function of the time after irradiation. The hypotheses considered here make different predictions for the shape of such curves. The theory of tissue-specific cell sensitivity to destruction postulates that the cells of each tissue have their own intrinsic susceptibility with respect to a radiation-induced process that causes removal of the cells. Since cells are also being removed by natural processes in this tissue, we shall call the postulated action of radiation one of accelerated removal. This theory can make no predictions about the dose at which observable depletion in the nucleated cells of the bone marrow will begin. It does predict, however, that as long as there are still cells left in the bone marrow, the rate of cell depletion will always increase steadily with increase in the radiation dose, since the larger the dose, the larger the proportion of the cell population that will receive the necessary number of hits for destruction. Only if there were some strange compensatory factors operating, could this behavior fail to be demanded by this theory.

The theory of reproductive inhibition, however, makes quite different demands and is also able to make some specific quantitative predictions. First, it is to be noted that the cell reproductive survival curve as shown in Figure 5-2 has a shoulder for the first 50 to 75 rads. Consequently, one would expect no visible cell depletion to be observed below that dose. A decrease in cell numbers is predicted beginning with doses above that point. Again, as the administered dose increases, one should observe steadily increasing rates of cell depletion since cell reproduction will be more progressively inhibited while cell removal continues at a constant rate. However, an important difference now occurs between the prediction of these two theories. According to the reproductive inhibition hypothesis, a dose eventually will be reached at which virtually all the normally reproducing cells in the bone marrow have lost their capacity for multiplication. At this point, further increase in dose of ionizing radiation should have no effect on the depletion of the nucleated cells of the bone marrow. This theory also predicts the magnitude of the limiting dose at which this final rate of cell depletion is reached. The single-cell survival curve of Figure 5-2 demonstrates that at approximately 250 rads, 90% of the cells have been killed reproductively. The uncertainty in measuring the number of nucleated cells remaining in the bone marrow lies in the neighborhood of 5 to 15%, largely because of unavoidable differences in normal bone marrow cell contents in different animals. Therefore, we would predict that for doses above 250 rads, the depletion curve for nucleated cells of the bone marrow should achieve a limiting slope that is not changed despite further increases in the administered dose.

The actual behavior observed in such an experiment carried out with young mice is shown in Table 5-2. In repeated experiments, no depletion occurs for doses below 70 rads. Beyond this point, depletion does occur, and it steadily increases with dose; in the neighbor-

Table 5-2 Nucleated Cell Content of Tibial Bone Marrow of Irradiated Mice (the mice were sacrificed 24 hours after irradiation)

Dose (rads)	*Nucleated Cell Count ($x\ 10^{-7}$)*
0	1.00 ± 0.20
50	0.88
100	0.52
150	0.38
200	0.28
250 - 1200	0.33

hood of 250 rads, the depletion rate achieved is maximal, remaining constant for doses up to 2000 rads.

Still another test of this theory is possible. If these animals are treated with a drug which is known to have only a single effect on cells, namely the prevention of mitosis, one would expect to obtain bone marrow cell-depletion curves that closely resemble those found with radiation. Specifically, the curve obtained with maximal doses of such a drug, should be the same as that obtained for doses greater than 250 rads of radiation. Studies with the drug colcemide carried out *in vitro* had shown that, for doses of 0.05 μg/ml or higher, mitosis becomes completely inhibited but no other stage of the cellular reproductive cycle is altered. Therefore, this drug was administered, in large concentrations, to a series of mice, and the number of nucleated cells in the bone marrow at various times was determined, just as had been done in the radiation case. Again, as predicted by the reproductive-inhibition theory, a depletion curve was obtained very closely resembling that obtained with limiting doses of radiation. It may be concluded that the nucleated cell population of the bone marrow behaves in accordance with the demands of the theory of cell reproduction inhibition and not at all in the fashion predicted by the theory of specific tissue cell destruction.

Some implications for mammalian physiology. These considerations explain the very old observation in mammalian radiobiology that the most rapidly dividing tissues are the ones which exhibit the greatest radiation pathology. As we have seen, this follows, not because these cells are particularly sensitive, but because these tissues have a very high normal rate of cell removal, which causes rapid depletion of the tissue when the compensating cell reproduction is eliminated by radiation. In addition, a variety of experiments have demonstrated the uniformly high sensitivity of the cell reproductive process as compared to other cell functions. The theoretical explanation for this lies in the fact that ionizing events occurring anywhere in the chromosomes can presumably contribute to cell reproductive death. On the other hand, events leading to specific differentiation or maturational damage, without affecting the ability of cells to reproduce, presumably can occur only at particular sites within the chromosomal structures and therefore would appear to offer a smaller target to the radiation.

The theory also explains for the first time, why the mean lethal dose of whole-body radiation for most mammals is approximately 400 rads. Since the survival curves of most mammalian cells are similar, this dose of irradiation produces approximately the same amount of killing in different cells. At 400 rads, approximately 95% of the reproducing somatic cell population has been killed, and this, then, would represent the point at which cell reproduction fails to compensate for the normal depletion by so large a margin as to cause death. These considerations also explain why, for doses in the vicinity of the near lethal range, death can be prevented if the animal is injected with bone marrow cells, which recolonize the depleted hematopoietic tissue. If other vital organs with rapid cell turnovers could also be recolonized by normal cells, perhaps radiation death from even higher doses could be prevented.

Similarly it now becomes possible to understand why antibody formation is inhibited by ionizing radiation. It has been demonstrated that, after administration of the antigen, antibody formation requires cell multiplication so that colonies of antibody-producing cells will be engendered. Radiation prevents this necessary step of cell reproduction and so can forestall the normal immune response.

An extremely interesting experiment, originally carried out by Patt, became readily explainable when the dynamic effects here described had been clarified. Working with frogs he found that the animals given high doses of ionizing radiation develop radiation sickness; but if their temperature was lowered directly after irradiation, the animals could be kept for long periods without the development of the typical symptoms of radiation sickness. When, however, their temperature was again elevated, the processes associated with the development of the typical radiation syndrome became instituted as though the radiation had only just been applied at the time of restoration of their body temperature. In accordance with the effects which we have been discussing, lowering the body temperature lowers the rate of cellular dynamic processes so that rates of cell removal from the various tissues are presumably slowed down. Hence the inability of the cells to divide does not hurt the animal appreciably at these low temperatures. Once the temperature is again elevated, normal removal processes are restored, and cell depletion and its consequences begin.

Perhaps one of the most dramatic confirmations of the validity of these ideas comes from radiobiological experiments with insects at different stages of their developmental cycle. The larval stage requires intense cell division. The adult stage, however, represents a maturational state in which little or no somatic cell division occurs. As would be expected then from these considerations, experiment has shown that the larva are indeed highly susceptible to killing by ionizing radiation, whereas adult insects can be exposed to doses of many thousands of rads with relatively minor pathological consequences.

5-6 HUMAN IMPLICATIONS

Availability of the survival curve for mammalian cells exposed to ionizing radiation clarified much that was obscure in mammalian radiobiology. The use of radiation in cancer therapy became a much more definitive tool when it was possible to estimate the cell survival to be expected after various doses. The mammalian radiation syndrome was enormously clarified, and the power of ionizing radiation as a research tool in many areas of mammalian biology was greatly expanded. However, biologists have come to understand that, because of the molecular biological nature of genetic structures, no exposure of living cells to ionizing radiation is without biological cost. Therefore, whenever ionizing radiation (or any other physical, chemical, or biological agent) is to be applied to a human population, it becomes necessary to define the extent of the deleterious effects which may be expected and then to weigh the expected gains against the biological cost.

The hazards due to small doses of irradiation, however, are especially difficult to evaluate, so that the actual cost becomes difficult and sometimes impossible to assess. The action of small doses is a probabilistic effect. Not all cells or animals which are irradiated with the same small dose will behave in the same way because of the randomness inherent in the basic action. In addition, many of the pathological effects of radiation require years, decades, or generations for expression. For example, if the gene mutations induced in the germ cells produce a heterozygous condition, it may be necessary to wait several generations until appropriate mating makes these lesions clearly recognizable. Some cancers produced by irradiation of somatic cells may take 20 years or more to develop. This long lag period is especially important in making extrapolations to man from experimental results on shorter-lived animals, such as the mouse and the rat.

In view of the various difficulties inherent in the definition of radiation damage in biological systems, how does one go about assessing the biological costs of any irradiation episode? First, it is necessary to make the best possible estimate of the probability of inducing each of the various lesions, such as different types of cancer, birth defects and other effects that may result from that given exposure; and, second, the cost of the lesion to the individual, to the family, and to society must be taken into account. Obviously, then, the biological cost depends not only on the dose, but also on the size of the exposed population. The probability for any given kind of damage to an individual may be as small as one in a million. However, if a population of 10 million people is to receive the irradiation, one can expect ten cases of the particular disease under consideration to result. If ten cases of Down's syndrome were to be included as the result of a given radiation experience, the cost to the affected individuals, their families, and to society would be enormous and cannot be measured by monetary considerations alone. It is necessary to make this estimate for each type of lesion and for every proposed exposure.

Further difficulties in accurately making such estimates are contributed by various factors:

(1) The existence of background radiation to which all humans are exposed may mask any extra damage that might result from comparable or lower exposures
(2) All of the aberrations which radiation can produce in the human genome can also occur spontaneously or through the action of other kinds of toxic agents
(3) Not all of the different pathological effects that can be produced by low doses of radiation have been delineated
(4) If the radiation is administered in the form of radioisotopes ingested by the body, many more uncertainties arise, due to the many uncertainties about the fate of such ingested materials in the body
(5) Little is yet known about how other agents may act to increase the damaging effects of radiation (caffeine, which inhibits radiation repair processes, might well change appreciably the amount of damage from a given radiation dose).

Because of these considerations, it is easy to see why the maximum dose that one may wish to consider giving to large populations (in return for very distinct advantages that will clearly outweigh the expected biological cost) will most likely be lowered as scientific study continues. This is due to the fact that, as we have studied radiation effects on mammals, we have become able to recognize more and more different kinds of pathological conditions produced by such exposures. This tendency is borne out by the history of the maximum permissible dose set up by various national and international agencies. This dose has steadily been lowered ever since its use was first proposed. Its magnitude declined by 1500% during the period between 1952 and 1959 alone, as shown in Table 5-3. The fact that the maximum allowable dose for large populations has decreased so steadily in recent decades should make one particularly cautious in accepting this quantity as a trustworthy figure. It should be emphasized that this concept of the maximum permissible dose is always based only on the methodologies existing at the time for demonstrating harmful effects due to radiation. Thus, any number which is considered at any period to be a "maximum permissible dose" is a poor substitute for the critical information which is actually needed, that is, a list of the numbers of cancers, birth defects, and other pathologies that result from exposure of a given population to a given dose.

*Table 5-3 History of the Change in the Maximum Permissible Dose, from 1952-1959**

Average Total Body Exposure† *(rem/week)*	*Date*
0.03	1952
0.01	1958 (April)
0.003	1958 (September)
0.002	1959 (May)

*K. Z. Morgan and J. E. Turner, *Principles of Radiation Protection*, Wiley, New York, 1967.

†By the National Committee on Radiation Protection, the International Committee on Radiation, or their advisory Committees.

The current value of the maximum allowable dose now needs to be carefully reconsidered with the object of decreasing this figure again by an appreciable factor, which might very well be as high as 10. The reason for this is that, since 1959, we have become aware of a whole new group of human diseases, apparently capable of being induced by radiation, whose genetic nature was unknown at the time the currently employed standards were adopted. These diseases constitute the genetic diseases due to chromosomal aberration. The true number of the human chromosomes was only discovered in 1956. The first complete identification of the human karyotype was carried out in 1958, and the first human disease due to chromosomal abnormality, Down's syndrome or mongolism, was identified very shortly thereafter. Since that time, the number of these diseases that are recognized in human newborns has steadily increased, and it is now apparent that approximately 0.5 to 1% of all human live births are accompanied by chromosomal abnormalities that may produce the most deep-seated disease in these babies. More than two-thirds of these include mental deficiency as one of the pathological symptoms. When one considers the enormous costs, to the affected families and to society, of caring for a mentally defective individual, it is evident that these diseases constitute one of the most serious public-health problems in this country. The theory of calculated risk demands that one consider both the probability of the undesired event and its cost. This set of diseases is so costly to society that a reexamination of the permissible dose of radiation for large populations must be carried out as soon as possible. In a field with so many uncertainties, and where error may have a tragically high human cost, it would seem reasonable to adopt policies which ensure that, if we err, it will be on the conservative side.

5-7 SUMMARY

The determination of single-cell survival curves defined the mean lethal dose of x rays for mammalian cells and demonstrated that ionizing radiation is far more potent in producing cell reproductive death than had been previously estimated. The survival curves *in vitro* and *in vivo* are very similar, and evidence has been collected demonstrating that the effective target for the lethal action is the chromosomal complement. The availability of the survival curve has clarified many aspects of the mammalian radiation syndrome, has afforded insight into the turnover rates of cells in normal mammalian tissues, and has illuminated important areas of radiotherapy of cancer. The genetic damage that radiation can exert on all living organisms makes it necessary, despite all its beneficial effects in medicine, to scrutinize human radiation exposures most carefully, particularly where large populations are involved.

REFERENCES

General

Elkind, M. M., and G. F. Whitmore. "The Radiobiology of Cultured Mammalian Cells," Gordon and Breach, New York, 1967.

Morgan, K. Z., and J. E. Turner. "Principles of Radiation Protection," Wiley, New York, 1967.

Puck, T. T. Cellular interpretation of aspects of the acute mammalian radiation syndrome," Symposium of the International Society for Cell Biology," vol. 3, Academic, New York, 1964.

Selected Papers

Hewitt, H. B., and C. W. Wilson. Observations concerning the eradication of mouse leukemia using whole body irradiation, *Brit. J. Radiol.* **31**, 340 (1958).

Patt, H. M., and M. N. Swift. Influence of temperature on the response of frogs to x-irradiation, *Am. J. Physiol.* **155**, 388 (1948).

Puck, T. T., and P. Marcus. Action of x-rays on mammalian cells, *J. Exptl. Med.* **103**, 653 (1956).

Russell, L. B. The effects of radiation on mammalian prenatal development, in "Radiation Biology," edited by A. Hollaender, vol. 1, McGraw-Hill, New York, 1954, p. 861.

Szybalski, W., and Z. Opara-Kubinska. DNA as principal determinant of cell radiosensitivity, *Radiation Res.* **14**, 508 (1961).

Tolmach, L., and P. Marcus. Development of x-ray induced giant HeLa cells, *J. Exptl. Cell Res.* **20**, 350 (1960).

CHAPTER 6

Some Molecular Aspects of Mammalian Cell Reproduction

The general pattern of mammalian cell biochemistry leaves no room for doubt that these cells obey the same fundamental principles as those which have been established for *E. coli.* The primary genetic template is DNA, as is the case with the simpler microorganisms. Enzymes, which will synthesize DNA with the aid of a preexisting DNA template and which will synthesize RNA from a DNA template, are isolatable from mammalian cells. Transfer RNAs are present in mammalian cells and, as in bacteria, multiple transfer RNAs exist for transport of individual amino acids. The amino acid activating enzymes have been demonstrated in these cells, and ribosomes and polyribosomes on which protein is assembled have been isolated. As in bacteria, RNA synthesis can be selectively inhibited by actinomycin and protein synthesis by puromycin (Figure 6-1). Thus, the general pathway DNA ⟶ RNA ⟶ protein clearly is as fundamental in mammalian cells as in bacteria.

Despite these basic similarities, however, the mammalian cell involves many complexities in the interrelationships of DNA, RNA, and protein dynamics, different from those of the bacterial cell. Delineation of details of these processes has barely begun. The purpose of this chapter is to indicate some directions currently under investigation and to describe tools that promise to produce large-scale illumination in this area.

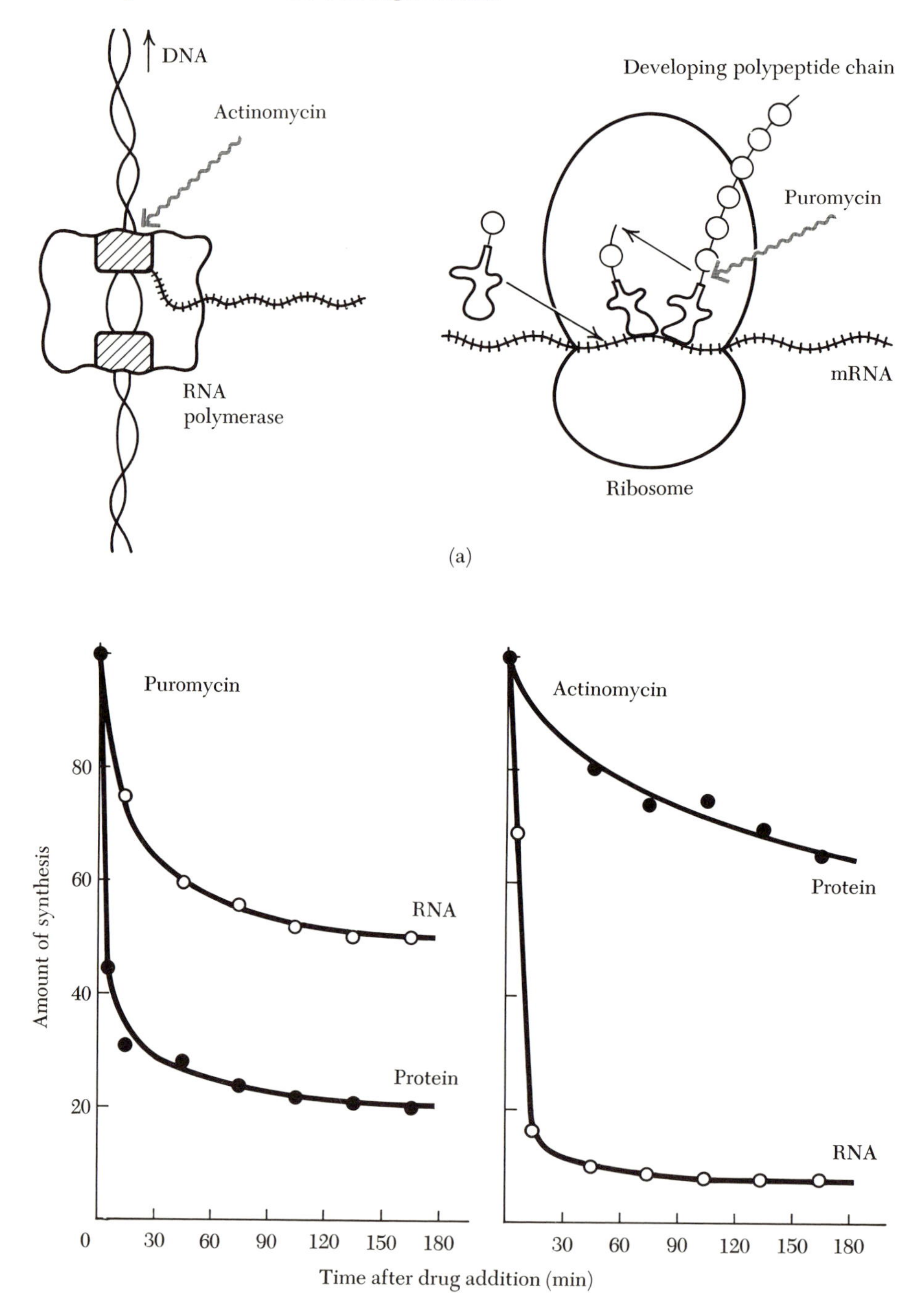
DNA
Actinomycin
RNA
polymerase
Developing polypeptide chain
Puromycin
mRNA
Ribosome
(a)
Puromycin
80
60
40
20
Amount of synthesis
RNA
Protein
Actinomycin
Protein
RNA
0
30
60
90
120
150
180
Time after drug addition (min)
(b)

Figure 6-1 (a) Schematic representation of the specificity of action of actinomycin and puromycin in blocking DNA and protein synthesis, respectively. Actinomycin, by forming a complex with the DNA, displaces RNA polymerase from the double helix and thus prevents the proper functioning of the enzyme. Puromycin, on the other hand, is itself incorporated into the growing peptide chain, with the result that incomplete polypeptides are terminated and released prematurely from the ribosome. (See Figure I-6 for details of protein synthesis.) (b) Effect of puromycin and actinomycin on RNA and protein synthesis of mammalian cells in culture. Results are expressed in terms of the amount of each synthesis relative to the control. Three points of interest emerge from this figure: (1) In each case there is a marked depression of synthesis corresponding to the primary target of the drug, followed by a slower drop in the second synthetic process. Thus, protein synthesis is necessary for the continued synthesis of RNA and, conversely, synthesis of messenger RNA must continue in order for new protein synthesis to occur. (2) The fact that puromycin does not drop the level of protein synthesis to zero reflects the ability of this drug to become incorporated into the growing peptide chain, causing the release of incomplete polypeptides. (3) Protein synthesis continues at relatively high levels long after actinomycin addition. This behavior is an indication of the greater stability of messenger RNAs in mammalian cells as compared with those of E. coli.

6-1 METHOD OF LIFE-CYCLE ANALYSIS AND SOME REPRESENTATIVE APPLICATIONS

In bacteria, DNA replication occurs throughout the entire life cycle. In the mammalian cell, however, the period of DNA synthesis is only one of four different subdivisions of the cycle. The path traced by a cell in the process of dividing is shown in Figure 6-2, a pattern first discovered in plant cells and later found to apply to the cells of most higher organisms.

One would like to be able to ask the following questions about the reproductive cycle:

(1) What are the necessary biochemical events in this cycle?
(2) To what extent is the order of these events necessary?[1]
(3) What differences, if any, exist in the reproductive life cycles of

[1]For example, it has been shown that cells starved for thymidine, which is needed for DNA synthesis, die much more rapidly when they are furnished with all of the amino acids needed for protein synthesis than if they are starved for one or more of these amino acids so that protein synthesis is prevented. This finding, which is similar to one known in bacteria, suggests that cell death can occur as a result of a dislocation between the order in which certain protein and DNA syntheses are carried out.

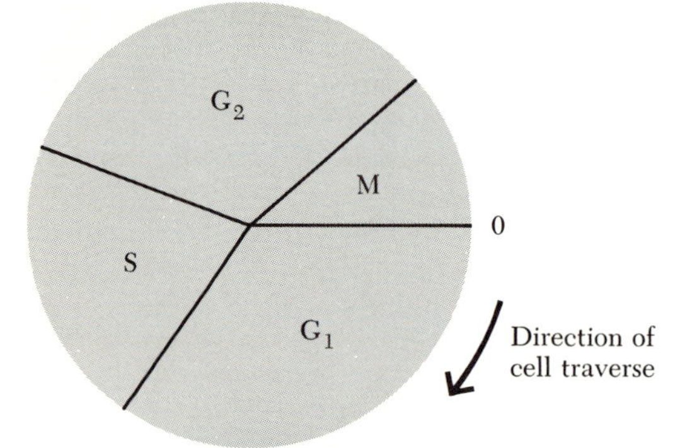

Figure 6-2 *Diagram of the major divisions of the reproductive life cycle of mammalian cells. G_1 is a period of intense protein synthesis and preparation for the DNA synthesis which occurs only in the region labeled S. In G_2 the doubled chromosomes shorten in length by a factor of 10,000 and become correspondingly thickened. As a result, the chromosomal threads, which would encompass a length of 2 meters in the fully extended condition, are compacted into short, thick rods that can easily be distributed equally among the two daughter cells, a process which occurs during mitosis, M.*

normal cells in different states of differentiation and in pathologic cells of different kinds?
(4) At what point in the life cycle do different physical, chemical or biological agents exert their influence?

Considerations of this kind made it seem worthwhile to develop methods for analysis of the biochemical events occurring at different parts of the reproductive cycle of the mammalian cell. A variety of different approaches have been used. One of the most direct is to phase the entire population in each of several positions in the life cycle and, then, to study the biochemical events which take place there. This procedure is discussed later (see page 142).

A more general approach which has proved valuable is the following. Consider a population of randomly dividing cells with a uniform generation time *T*. It is assumed that a specific inhibitor is available which, when added to the culture, stops the cell traverse of the life cycle at one particular point only. With the passage of time, the cells will accumulate in that stage of the life cycle just prior to the block. For simplicity we shall assume that this blocking point occurs at the end of mitosis, so that cell division is not completed and cells are accumulated in the mitotic phase of the life cycle (Figure 6-3).

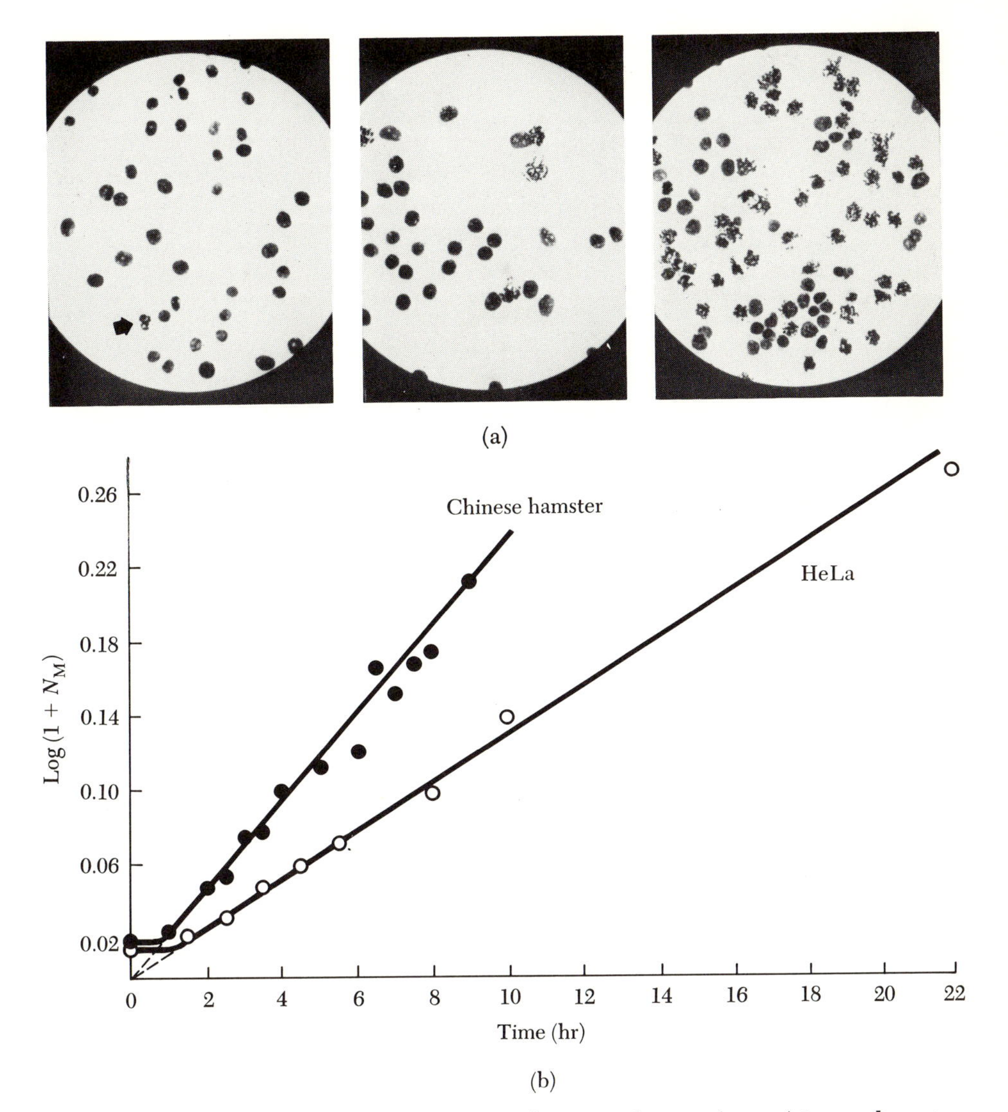

Figure 6-3 *(a) Typical accumulation of mitotic figures (arrow) in a culture to which an inhibitor of mitosis has been added. The inhibited cells remain in mitosis and their number steadily increases with time as more cells enter this phase. (b) Diagram of the behavior of the accumulation function, log (1 +* N_M*), where* N_M *is the fraction of mitotic cells in a culture treated with a mitotic inhibitor. This relationship permits analysis of various biochemical events occurring during the course of the reproductive life cycle and permits identification of the points in the life cycle at which various drugs act. Accumulation functions for two different cells are shown: a Chinese hamster lung cell and a human cancer cell. The latter reproduces more slowly than the former.*

Under these circumstances, it can be shown that the accumulation of cells in mitosis can be represented by the equation:

$$\log (1 + N_M) = 0.301\,(T_M + t)/T \tag{1}$$

where N_M is the fraction of the cell population in mitosis at any time, T is the average generation time of the culture, T_M is the time required for mitosis, and t is the time since the addition of the blocking agent. The quantity $\log (1 + N_M)$ is called the accumulation function (see derivation in Appendix V). Equation (1) predicts that, for a random culture, the accumulation function will consist of a straight line, from whose slope the generation time T can be calculated. If all the cells have exactly the same value of T, the accumulation

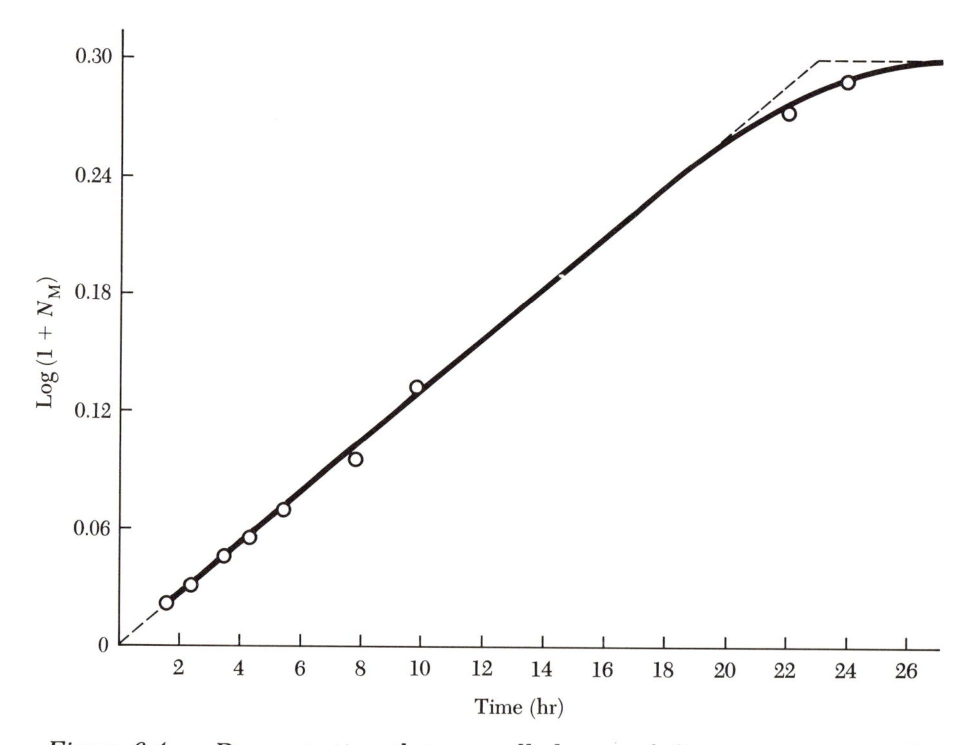

Figure 6-4 *Demonstration that a small degree of dispersion exists in the generation times of different cells in the same clonal culture. If T were exactly the same for every cell, the accumulation function would then abruptly become constant as shown by the dotted line above. The experimental points, however, indicate a slight lag of the slower moving cells in entering the mitotic phase. In the HeLa cell, the standard deviation in the generation time T was found to be ±9%.*

function would be linear until all of the cells were in mitosis, at which time the curve would become horizontal, with a discontinuity as shown in the dotted line of Figure 6-4. If T is not absolutely constant, but has a small variance, the collection function will curve in a gradual fashion to meet the horizontal asymptote. The experimental points in Figure 6-4 indicate the latter to be the case.

Experiments were carried out with the agent colcemide, which is known to inhibit cells in mitosis, to determine whether its effect on cells would correspond to the theoretical behavior shown in Figure 6-5. Instead of the behavior shown in the top of the curve, that of the

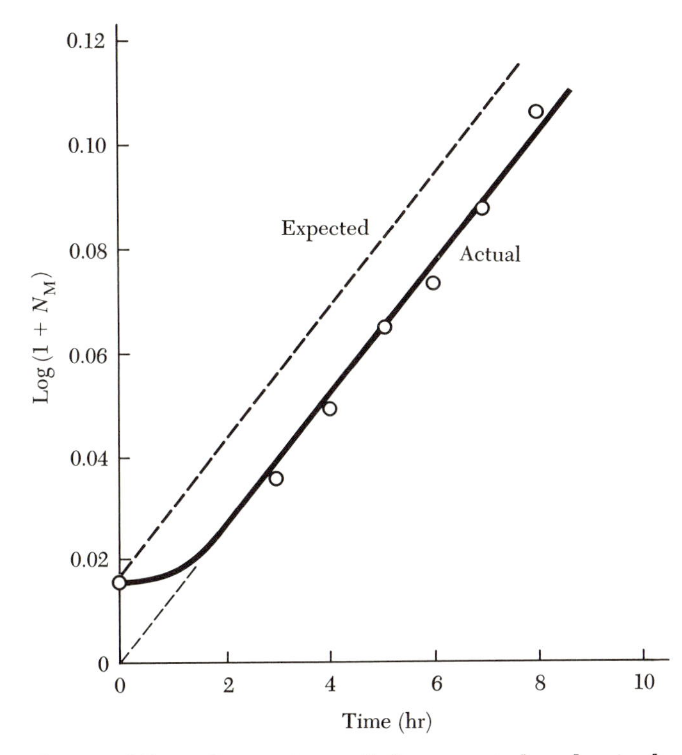

Figure 6-5 Comparison of the expected and actual curves obtained when colcemide was added to a HeLa culture and periodic sampling carried out for determination of the accumulation function. If the cells already in mitosis at the time of colcemide addition had been retained at that point by the drug, the curve should have started up linearly from the Y-axis intercept. Instead, there is an initial lag of slightly over an hour in the collection of mitoses, and the final curve is a straight line that extrapolates back to the origin. This behavior indicates that only the cells entering mitosis after the addition of colcemide are blocked by the drug.

lower curve was obtained. The generation time calculated from the slope of the experimental curve was 20.0 hours, as contrasted with the experimental value of 20.2 hours obtained on the same culture by means of time-lapse photomicrography. The horizontal displacement of the two curves has the following explanation: those cells which are already in mitosis at the time of drug addition are not inhibited but, rather, complete mitosis and again enter interphase in normal fashion. However, all new cells entering mitosis are blocked and accumulate. Thus, the curve obtained has the theoretically predicted slope, but it is shifted to the right by an amount equal to T_M, the time of mitosis. The actual equation then is as follows:

$$\log (1 + N_M) - 0.301\, t/T \qquad (2)$$

(see Appendix V).

Experiments of this kind demonstrate that the drug colcemide, which is known to inhibit the formation of microtubules, has no effect on the rate of progression of cells through their reproductive life cycle until the point is reached at which mitosis begins. Cells in the presence of the drug may enter mitosis, as demonstrated by the formation of the characteristic condensed state of the chromosomes, but they cannot leave it, and so mitotic figures accumulate. Hence it may be deduced that microtubule formation plays no important role in regulating cell reproduction until the time arrives for formation of the mitotic spindle.

The effectiveness of the accumulation function can be extended by the simultaneous addition of both colcemide and tritiated thymidine to a culture. One can then construct the ordinary collection function, first, for total mitoses, a second for those mitoses that are labeled with tritiated thymidine, and a third for total cells that have incorporated the tritiated thymidine into DNA. The three kinds of collection functions for a typical cell are shown in Figure 6-6. By

Table 6-1 The Relative Durations of the Different Phases of the Life Cycle in HeLa Cells, Compared to Those of the Chinese Hamster Ovary Cells in the Same Medium

	T (= Total Generation Time) (hr)	T_M/T	T_{G_2}/T	T_S/T	T_{G_1}/T
HeLa	20.1	0.055	0.22	0.29	0.41
CHO	12.4	0.065	0.21	0.33	0.39

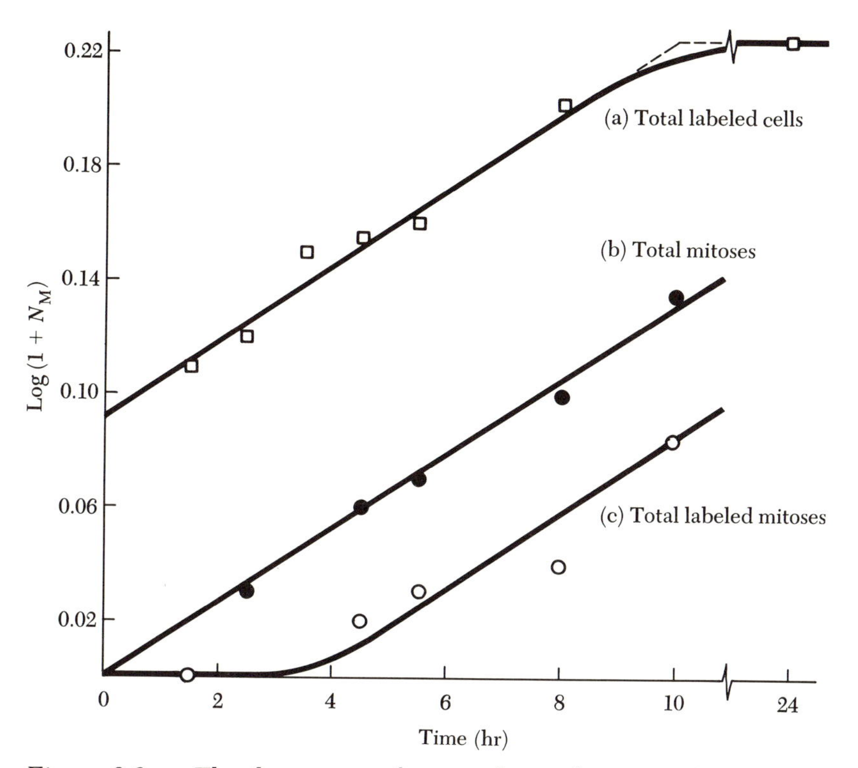

Figure 6-6 *The three types of accumulation functions that can be obtained when colcemide and tritiated thymidine are added simultaneously to a cell culture. The duration of M and G_2 can be determined by analysis of curves (b) and (c), while T_S and T_{G_1} are obtainable from (a).*

means of such data, the duration of each of the four major parts of the life cycle can be accurately calculated. A comparison of the values obtained for the Chinese hamster ovary cell and the HeLa cell, which have very large differences in their overall generation time, is shown in Table 6-1. It is of interest that, despite this large difference in generation time, the proportion of the total occupied by each of the four major parts of the life cycle, is reasonably constant.

Two other drugs, vincristine and vinblastine, were found to give accumulation functions identical to that of colcemide, thus furnishing evidence that these drugs function in the same fashion.[2] Drugs which

[2]These considerations apply only when the concentrations of vincristine and vinblastine are just sufficient to cause mitotic arrest. If higher concentrations are employed, inhibitory effects may also be found in other portions of the life cycle.

affect other parts of the life cycle can be similarly analyzed by this technique.

Determination of randomness of a culture. One of the most immediate purposes to which this methodology can be applied is the determination of whether or not a given population of reproducing cells is a random culture. A random culture can be defined as one which will produce a collection function, such as that shown in Figure 6-4, in the presence of an appropriate mitotic inhibitor. Any deviation from randomness introduces phasing into the cell population, and the resulting collection function will not be a straight line, but will show characteristic departures from this behavior. These deviations from linearity make it possible to deduce the distribution of cells along the life cycle in the initial culture.

Mode of action of other drugs and agents. The action of thymidine, a compound interesting because it is required by the Chinese hamster cell for DNA synthesis yet, in high concentration, is inhibitory to cell growth, has been analyzed by this method. The action of this agent is shown in Figure 6-7. The addition of an excess of thymidine, along with colcemide, produces a collection function normal only for the duration of the G_2 period. Therefore, the thymidine prevents cells from finishing S. Moreover, when thymidine is washed out of the culture, the collection function which is then obtained demonstrates that all of the cells in S had been stopped in their progression when thymidine was added. Therefore, this agent appears to halt cell progress throughout the entire S period.

A second example involves analysis of the reversible lag produced in mammalian cells, as in other cells, by the action of small nonkilling doses of x irradiation. Figure 6-8 shows the collection function obtained by an x irradiation of only 9 rads. This dose does not kill any of the cells but temporarily inhibits the cell contingent in the central region of the G_2 period. During the first 0.70 hours, the collection function is normal, demonstrating that the cells in this interval have not been impeded at all. Thereafter the curve is flat for the subsequent 0.70 hours, indicating complete cessation of progression toward mitosis. Then the curve again rises and, for 0.65 hours thereafter, the slope is virtually the same as in the control, a fact taken to indicate that the cells on the other side of the inhibited population are now coming into mitosis without any delay. From that point on, until the 3-hour point, the slope is greater than normal, indicating that, during this period, cells that have not been held up are being rejoined by those that have recovered from the lag, so producing a slope greater than normal. The fact that, after the 3-hour point, the radiation-delay curve and the normal curve coincide demonstrates the completely reversible nature of this particular growth inhibition.

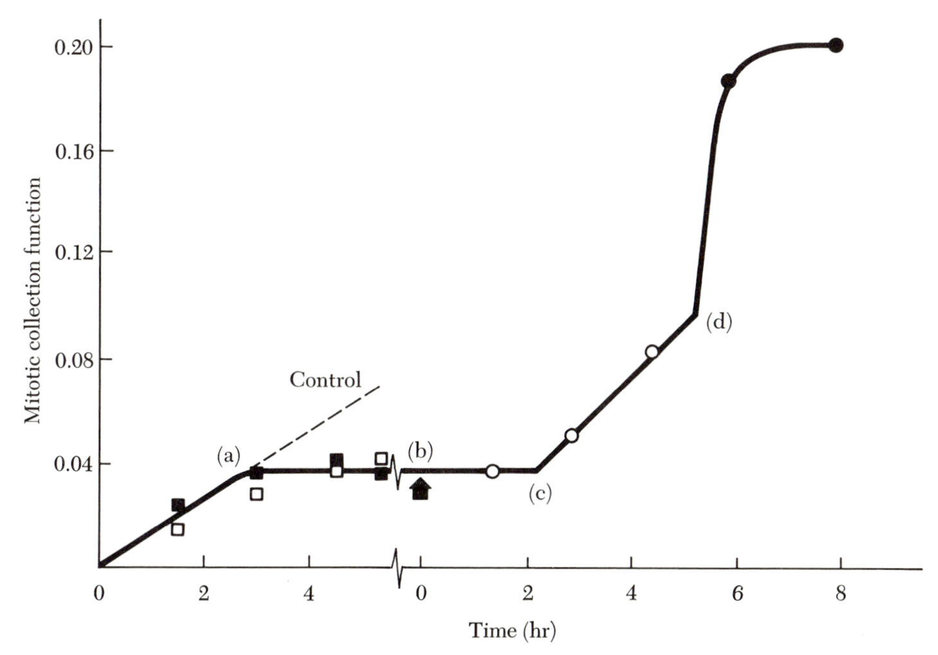

Figure 6-7. Determination of the mode of action of thymidine on the life cycle of HeLa cells. At zero time, an excess of thymidine is added simultaneously with colcemide, and cells are allowed to accumulate in mitosis. The collection function is normal for about 4.5 hours (a), after which it levels off. Since the length of the G_2 period in these cells is 4.5 hours, it would appear that all of the cells in G_2 at the time of thymidine addition enter mitosis normally, but that cells originally in S have been blocked. If, after a period, the thymidine is washed out (b), no cells arrive in mitosis for the first 4 to 5 hours thereafter (b)-(c), which is the time needed for the cells released from the block to traverse the G_2 period. Then, mitotic collection begins again and proceeds at an approximately normal rate for about 6 hours (c)-(d), which is the normal duration of S. At the end of this period, there is a sudden burst of mitoses, which encompasses virtually all of the remaining cells. Thus, the excess thymidine results in a cell distribution in which the S period is approximately that of a random culture, but all the rest of the cells accumulate at the end of G_1. Removal of the thymidine then causes these cells to proceed around the cycle at a rate approximating their normal progression.

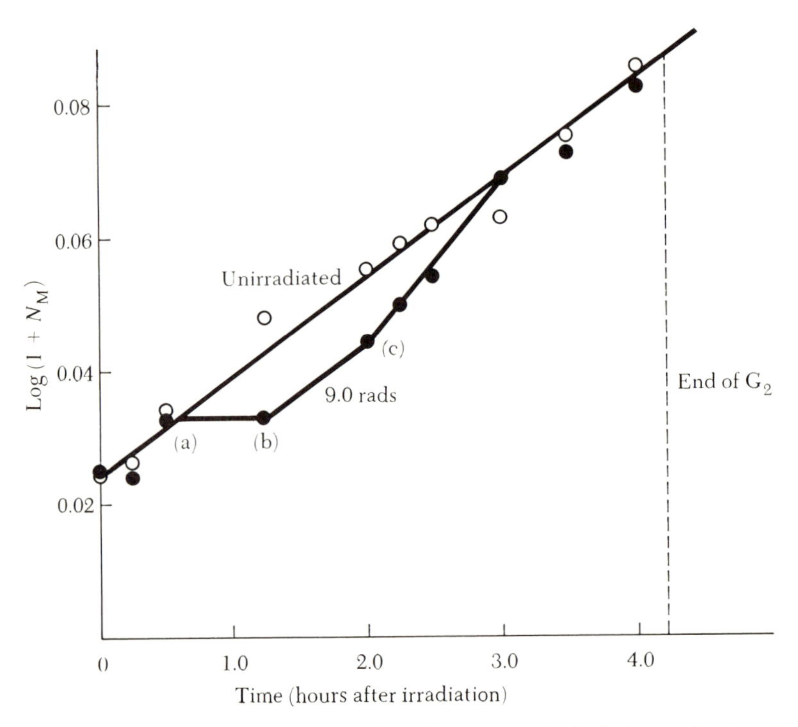

Figure 6-8 *The change produced by a nonlethal dose of x irradiation in the mitotic accumulation function of HeLa cells. Analysis of the data by the life-cycle methodology described in the text reveals (1) cells in G_2, beginning at 0.6 hours before mitosis (point (a)) and extending back until 1.3 hours before mitosis (point (b)) are delayed in reaching mitosis; (2) cells on either side of this 0.7-hour region are not delayed appreciably; (3) the magnitude of the "effective" delay (distance of (a) to (c)) is 1.4 hours; (4) the delay is reversible. This type of life-cycle analysis thus offers a means for the gaining of understanding of the action of many physical and chemical agents on mammalian cells.*

Phased cultures. Life-cycle analysis is used to determine the position in the life cycle at which various biochemical events occur. A related problem is the production of phased cultures, that is, cultures in which the cell population is not randomly distributed but is concentrated within a fairly narrow range of the life cycle. While phased cultures can be used to answer some of the problems for which life-cycle analysis is designed, it is more suited as a preparative rather than an analytical method. By phasing cells within a fairly narrow region of the life cycle, it becomes possible to extract from them particular enzymes, structural proteins, and other metabolites that might be present at only one point in the life cycle and thus maximize the yields of specific biological activities. A number of different procedures have been proposed for obtaining phased cultures. One of these relies on the fact that cells in mitosis round up and are therefore attached to monolayer surfaces only over a small area of the cell's surface. Such mitotic cells can easily be shaken off to furnish a suspension of cells, of which the great majority are phased in mitosis. Another method depends on the fact that agents like excess thymidine prevent DNA synthesis throughout the period of S. Since 28% of a random population of HeLa cells is distributed throughout the entire length of the DNA synthetic region S, no agent which stops all S cells can produce a strongly phased culture. However, by applying an agent like thymidine long enough to accumulate all the cells in S, then relieving the block long enough for all the cells to finish S, and, finally, re-adding an inhibitory concentration of thymidine, the cells can all be accumulated in a sharply defined interval at the beginning of S, as shown in Figure 6-9.

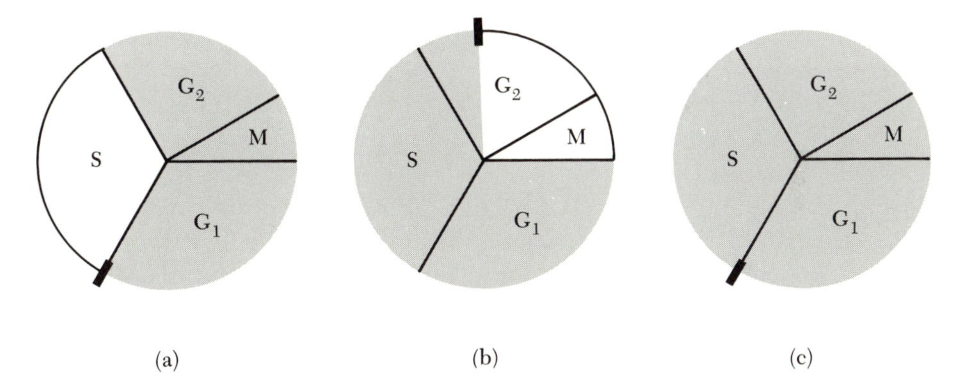

Figure 6-9 *Diagram illustrating the procedure by which cells in culture can be synchronized by the double thymidine-phasing method. After the initial addition of thymidine, cells in S are immediately arrested, while those in G_2, M, and G_1 continue through the cycle until they reach the beginning of S where they are also blocked (a). Thymidine is then washed out, and the cells are allowed to continue development until all have passed out of S into M and G_2 (b). The second addition of thymidine then results in the accumulation of all cells at the entrance to S (c). Thymidine can again be removed and the phased culture allowed to travel to any desired point in the cycle.*

6-2 PROCESSES LIMITED TO SPECIFIC DIVISIONS OF THE LIFE CYCLE

The life cycle of the mammalian cell consists of successive states, each of which accomplishes a necessary part of the overall process at a particular time. The delineation of the molecular characteristics of these different states, their necessary order of precedence, the degree to which particular steps can be altered with respect to position, direction, or relative rates in the cycle, and the pathological consequences of various toxic agents and disease states have hardly begun to be delineated. An enormously rewarding area of research is available here for the adventuresome experimentalist.

The G_1 phase. The newborn cell emerging from mitosis finds itself in the G_1 phase (Figure 6-2). The nuclear membrane and the nucleoli, both of which disappear in mitosis, are present again in G_1, and protein and RNA syntheses, which had been sharply depressed in mitosis, are resumed at a rapid rate. Particular enzymes have been shown to be specifically synthesized in G_1. Thus, ribonucleotide reductase and thymidine kinase, both of which are needed for DNA synthesis, are found to be at low levels throughout G_1 until just before DNA synthesis begins in S. The activities of these enzymes then increase and remain at high levels in S, but drop again to their original low levels in mitosis. Thus, it appears that, when a given enzyme has discharged its function in the life cycle, it may be selectively destroyed or inactivated. The dynamics of proteolytic and other inactivating actions on intracellular proteins and the triggering stimuli that bring them into operation are still obscure. However, available evidence hints that the regulatory mechanisms of the mammalian cell utilize proteolytic or other enzyme-inactivating processes, as well as control of protein synthesis, in order to achieve the necessary balance required by the complexity of the system.

The G_1 period, presumably, also serves as the point at which many proteins associated with particular differentiation functions are synthesized. It is of interest in this connection to note that in at least one cancer cell, the mouse ascites tumor, the G_1 period is virtually nonexistent. Since this cell reproduces effectively, despite loss of many of the biochemical steps ordinarily associated with the G_1 period, it seems a plausible inference that many of the processes ordinarily contributing to the length of G_1 are connected with differentiation rather than reproduction.

Some investigators have coined the name G_0 as a state in which cells may be arrested after completion of mitosis but prior to initiation of the specific molecular events of G_1. The utility of this definition must await more detailed analysis of the biochemical events involved in the transition from mitosis to initiation of DNA synthesis.

In G_1, the chromosomes are released from the condensed mitotic state and are no longer recognizable as distinct entities. Instead, the nucleus appears as a fairly homogeneous body. Presumably, this uncoiling of the chromosomes exposes to RNA polymerase the specific regions of the genome that are to be active in RNA synthesis. The 250 Å chromatin fiber, which exists in mitotic chromosomes (see Chapter 2), has also been demonstrated to exist in interphase chromosomes.

The S period. It has been customary to associate mitosis with reproduction in mammalian cells. In actual fact, however, the mitotic process is the last part of the division cycle in which the products of reproduction furnished by preceding steps are simply distributed among the two daughters. The heart of biological reproduction takes place in S when the chromosomes and the genes they bear are duplicated.

In *E. coli,* there is only one point at which initiation of DNA synthesis can occur, and therefore synthesis is begun at one end and continues linearly until the entire chromosome has been replicated. The mammalian cell chromosomes have many initiation points. This arrangement permits a high degree of regulation in the order of gene replication. The interval between two successive initiation points on a chromosome is called a replicon. It has been estimated that the replicon is between 7 and 30 microns long and that there are approximately 50,000 of these units in the mammalian genome.

Table 6-2 summarizes some of the vital statistics of DNA in *E. coli* and in mammalian cells. It is evident that the rate of DNA synthesis per replicon is much slower in the mammalian cell than in *E. coli.* Part of the reason may lie in the fact that the mammalian cell's growing points are not all simultaneously active, but require a specific triggering action to be initiated. Evidence for such a communication system between different parts of the DNA replicatory system is already known to exist in mammalian cells. Thus the second X chromosome which forms the Barr body in female cells and the Y chromosome in male cells are always the last to be replicated, while certain other chromosome pairs tend, largely, to be replicated early in S. The molecular nature of information transfer between chromosomes is completely unknown, as is the physical nature of the chromosomal association with the replicating apparatus. It is of interest in this latter connection, however, that evidence for chromosomal attachment to the nuclear membrane, suggestive of that previously described in bacteria, has recently been found. Thus, if mammalian cells are centrifuged at velocities varying between 20,000 to 300,000 g, the chromatin material in the nucleus becomes piled up in a small part of the nucleus attached to the nuclear membrane. Electron micrographs also appear to demonstrate that DNA synthesis may be initiated at the nuclear membrane.

Table 6-2 Vital Statistics for DNA

	Mammalian Cell (Human Liver)	*E. coli*
Total amount (g/cell)	2×10^{-11}	1×10^{-14}
Total amount in mitochondria (g/cell)	3×10^{-14}	–
Average number of replicons per cell	5×10^{4}	1
Average rate of synthesis (nucleotide pairs/replicon/sec)	20	5×10^{3}

The DNA double helix appears to be associated with protein in the form of fibers, in which, at least for part of the life cycle, the DNA is continuous. This is shown by hydrolytic experiments in which continuity of these fibers is destroyed by DNAse but not by proteases, such as trypsin, even though these latter agents somewhat reduce the fiber thickness (see Chapter 2). The fact that DNA replication is initiated simultaneously at many points along each chromosome might require the production of single-stranded breaks at each initiation site. Presumably, a special enzyme would carry out this action, and these initiation sites, then, would contain the same sequence of DNA codons that such an enzyme would recognize. Repetition of a common DNA sequence at the beginning of each replicon might be involved. Preliminary evidence for the existence of single-stranded breaks in the DNA chain at particular points in the life cycle has been reported, but these studies are still in an early stage.

The G_2 period. The G_2 period bridges the interval between the end of DNA synthesis and the initiation of mitosis. In the S period, the replication of each segment of DNA requires its maximum unfolding, while, in mitosis, the chromosomes have become discrete bodies thick enough to be readily visible in the light microscope. Thus condensation of the chromosomal structures presumably begins sometime in G_2. Calculation shows that the human chromosomes have been condensed approximately 10,000 times with a concomitant increase in their thickness by the time they are in prophase. This process occurs through a supercoiling phenomenon involving complexing of histones with DNA and subsequent folding up of the fibers, so that further and further degrees of supercoiling occur. The process of chromosomal condensation is highly specific. The final length of each arm of each chromosome is constant in any species and, indeed, permits identification of the individual chromosomes as discussed in Chapter 2. Moreover the thickness of the condensed chromosomes is reasonably constant within any cell, indicating that the process of condensation is fairly uniform among the different chromosomes.

The enormous decrease in chromosomal length, which occurs in G_2, is necessary for the subsequent equal division of the chromosomes in the two daughter cells, which occurs in mitosis. The sequence of physical and chemical changes underlying the rapid assumption of the ordered structure of the condensed chromosome constitutes one of the most interesting examples of ordered motion on the molecular scale that living forms have learned to carry out.

The ease of observation of the transition from the G_2 period to mitosis makes a number of simple measurements possible. For example, the G_2 period of the Chinese hamster ovary cell *in vitro* is approximately 3.1 hours. When actinomycin is added to a culture, RNA synthesis is inhibited almost immediately but cells continue to enter mitosis for an additional 1.3 hours. If puromycin is added instead, protein synthesis is inhibited almost immediately and so is the entry of cells into mitosis. This experiment indicates that the last synthesis of messenger RNA necessary for the initiation of mitosis occurs 1.3 hours before its associated protein synthesis. The production of a reversible lag in G_2, as a result of administration of low doses of ionizing radiation, has been described in connection with Figure 6-8. The mechanism of this lag is unknown.

Mitosis. Mitosis is the final step of the cellular reproductive process; it culminates in the birth of the new daughter cells. It is unique to the cells of higher organisms. A simple cell with a single chromosome would need only primitive apparatus in order to distribute equally, to the daughter cells, the two chromosomes resulting from DNA replication. A cell with 46 chromosomes, on the other hand, will

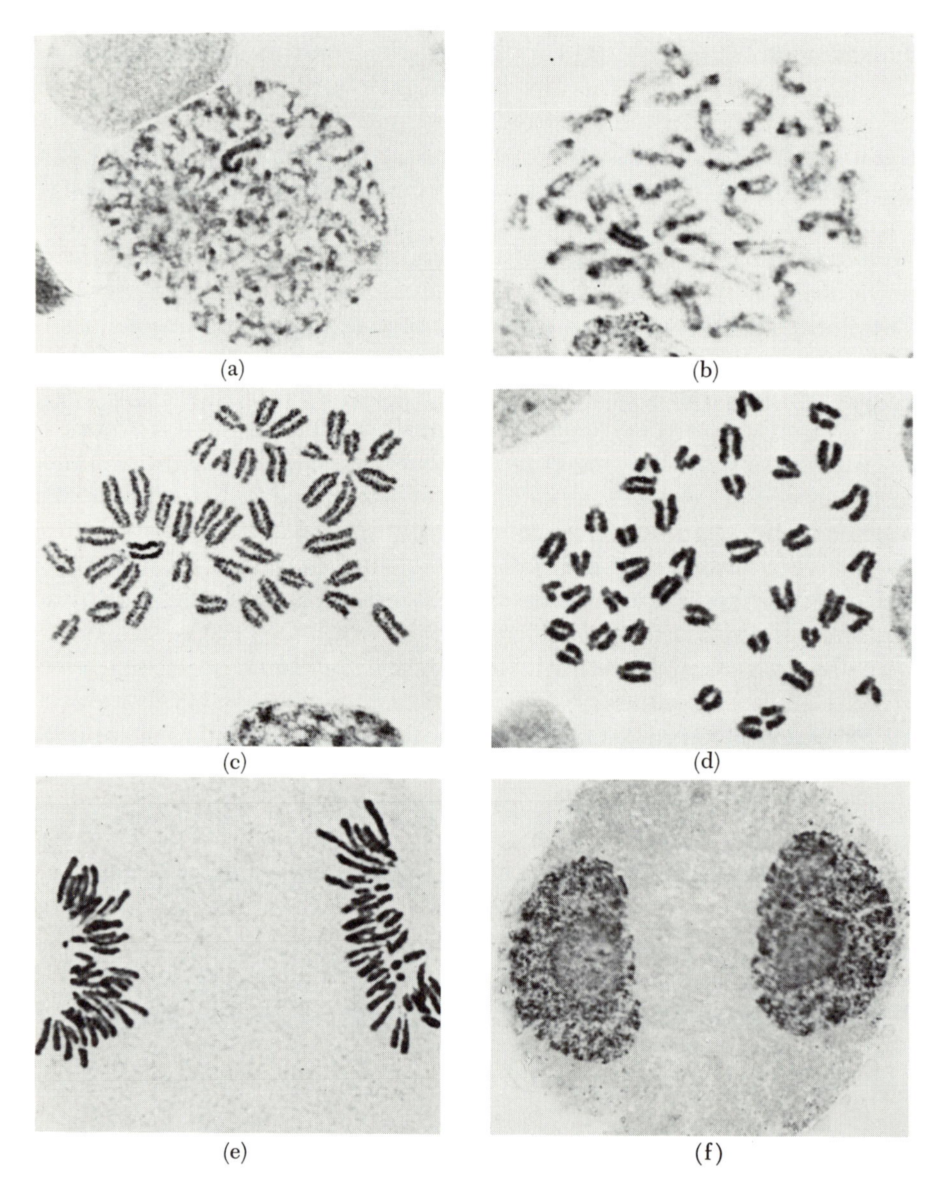

have 92 such bodies after replication, so a complex set of structures and operations is necessary in order for each daughter cell to receive an exact set.

Mitosis is heralded by a rapid rounding up of the cell, disappearance of the nuclear membrane, and the appearance of discrete, condensed chromosomes. At the same time new antigenic materials appear in the cell surface. The centrioles migrate to opposite poles of the cell. Protein fibrils are then formed, radiating from the centrioles to the cell membrane, on the one hand, and to the centromere of each chromosome, on the other, thus forming the spindle apparatus. The members of each chromosome pair are then pulled to opposite poles of the cell. The spindle fibrils consist of a single kind of protein, organized into unbranched, rigid structures called microtubules. These can be depolymerized into monomeric units of a protein with a molecular weight of 60,000. Colchicine, an alkaloid derived from the crocus *Colchicum autumnale*, specifically binds to microtubule subunits, prevents their organization into larger, functional structures, and therefore blocks mitosis in cells of higher organisms. Bacteria, which lack the spindle microtubules, are unaffected by this drug.

When the chromosomes have been distributed to the poles of the cell, the cellular membrane becomes extended, so as to enclose each of the new daughter cells. The remaining processes of mitosis are concerned with returning these cells to the fully functional state. Thus, the nuclear membrane and the nucleoli are reformed and the chromosomal condensation processes are reversed, so that gene products can again be synthesized. Pictures illustrating the four well-known parts of mitosis are presented in Figure 6-10.

There is some evidence that the reduction in protein synthesis during mitosis may be due to the temporary inactivation of the ribosomes during this period. The lack of RNA synthesis is not difficult to

Figure 6-10 Photomicrographs of mouse chromosomes in various stages of mitosis. (a), (b) Prophase, which is characterized by shortening and thickening of the chromosomes to the point where they can be seen as discrete entities. (c) Early metaphase (or prometaphase), in which the condensation process continues. (d) Metaphase, during which the spindle apparatus is formed. (e) Anaphase, during which the centromeres have attached to the spindle and the homologous chromatids move to opposite poles of the cell. (f) Telophase, in which the chromosomal condensation, which has reached a maximum in anaphase, now reverses. The nuclear membrane is reformed and the nucleoli begin to reappear. (Photographs courtesy of J. H. Tjio.)

rationalize since, by this time, the chromosomal template has become so highly condensed that exposure of the DNA sequences to the medium is strongly reduced.

In contrast to the synthesis of RNA in the nucleus, the RNA in mitochondria is synthesized at a normal rate throughout mitosis. This synthesis appears to be different in kind from that of nuclear RNA in that the main products appear to have slightly different sedimentation constants, and the synthesis is sensitive to ethidium bromide; nuclear RNA synthesis, on the other hand, is more sensitive to ultraviolet irradiation and virus infection.

The fiber structure of the mitotic chromosomes, as revealed by electron microscopy, was discussed in Chapter 2. These fibers appear to travel both longitudinally and transversely along and between the two chromatids in the fixed preparations. As mitosis proceeds, the number of transverse fibers decreases until, finally, the two sister chromatids separate. As discussed in Chapter 2, there is evidence that the DNA in the region of the centromere of each chromosome has similar base sequences. Presumably, therefore, these regions are involved directly or indirectly, that is, through mediation of specific RNA or protein, with the attachment of the spindle fibers. It is of interest to compare the precision of the mechanical movements attained by the cell in mitosis with that familiar in man-made machinery. The latter almost always involves much larger parts, most of which are rigid. The biological mechanism achieves energy conversion and highly ordered motions using structures that are much smaller and flexible. When these processes become better understood, they might easily provide a new model for man-made mechanisms.

Interaction between different phases of the cellular life cycle. Availability of the cell-fusion technique makes it possible to combine cells from different phases of the life cycle into a single composite cell, so that the effects of the interaction can be studied. Thus, HeLa cells have been phased at particular points in their reproductive cycle and then fused with cells phased in some other period. For example, HeLa cells, phased in mitosis, were fused with other HeLa cells phased in G_1, S, or G_2. The two different cells could be distinguished by labeling the nuclei of one set with tritiated thymidine before fusion. Within 15 minutes after fusion with the mitotic cells, condensed chromosomes appeared in the nuclei of the interphase cells. When G_2 cells are used for this fusion, the newly condensed chromosomes are doubled and roughly resemble mitotic chromosomes. When G_1 cells are used, however, the condensed chromosomes are single, as would be expected from cells that had not yet replicated their DNA. When cells in the S period are fused with these mitotic cells, chromosomal condensation occurs, but this is also accompanied by extensive "pulverization," that is, apparent discontinuity in the condensed chromatin (Figure 6-11). It is conceivable that these apparent discontinuities arise at points where DNA synthesis is taking place in individual replicons. Introduction of the breaks in the DNA chain, as synthesis is initiated, may result in interruption of the resulting condensation of the chromosome when fusion with a mitotic cell occurs.

These experiments indicate that mitotic cells contain chemical factors that can induce condensation of chromosomes from nonmitotic cells. These inducing factors operate over a broad range of species. For example, mitotic HeLa cells can induce chromosomal condensation in tissue-culture cells of the Chinese hamster or mouse taken directly from tissue biopsies. Indeed, it has been possible by this method to induce chromosomal condensation in bull sperm. The chemical elucidation of the structure of these chromosomal condensing factors and the dynamics of their synthesis and removal in the cell promise to be of enormous interest.

Other factors responsible for development of other phases of the life cycle of mammalian cells have also been demonstrated. Thus, when cells phased in S are fused with cells phased in G_1, the G_1 nuclei initiate early DNA synthesis. The most direct interpretation of this experiment is that some material, which triggers the initiation of DNA replication, is transferred from the cytoplasm of the more advanced cell to the chromosomes of the less advanced one.

6-3 PROCESSES OCCURRING THROUGHOUT LARGE REGIONS OF THE LIFE CYCLE

RNA metabolism. RNA governs all protein manufacture. It is synthesized throughout all parts of the life cycle (except mitosis) and consists of (1) the messenger RNA, which carries the genetic information for translation of the specific protein structures, (2) the transfer RNA, which carries the amino acids to their appropriate positions in the growing polypeptide chain, and (3) the ribosomal RNA, which

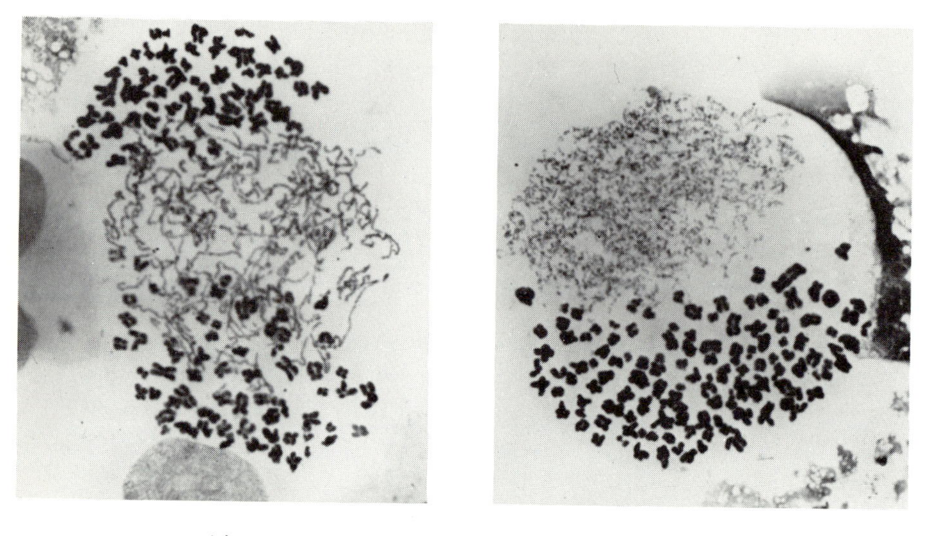

(a) (b)

Figure 6-11 Appearance of the chromosomes in hybrid cells formed by the fusion of two HeLa cells phased in different parts of the life cycle. (a) Fusion between a cell in mitosis and one in G_1. Although the chromosomes of the G_1 cell did not replicate, they were induced to condense by the presence of an inducing factor in the cytoplasm of the mitotic cell. Thus the G_1 chromosomes appear as single strands of chromatin rather than the normal duplicated state. (b) Fusion between a cell in mitosis and one in S. The chromosomes from the S cell had begun replication but were induced to condense before the process had been completed. The premature condensation appears to have been accompanied by extensive fragmentation and pulverization. (c) Fusion between a cell in mitosis and one in G_2. Since the chromosomes of the G_2 cell had already completed DNA replication, their morphology upon condensation, resembles that of the normal doubled mitotic chromosomes.

makes up a large part of the ribosomes on which the polypeptide chains are assembled. Mammalian cells also produce other kinds of RNA whose function is still unclear. Understanding of RNA dynamics in mammalian cells is still fragmentary.

The RNA metabolism of mammalian cells is more complex than that of bacteria. For example, there exists a species of mammalian RNA that appears to be contained in the nucleus and never reaches the cytoplasm. This material is highly heterogeneous, with sedimentation constants varying between 60 and 70S, and undergoes continuous and rapid synthesis and degradation in the nucleus. Approximately 50% of such material is degraded within a matter of seconds after its synthes-

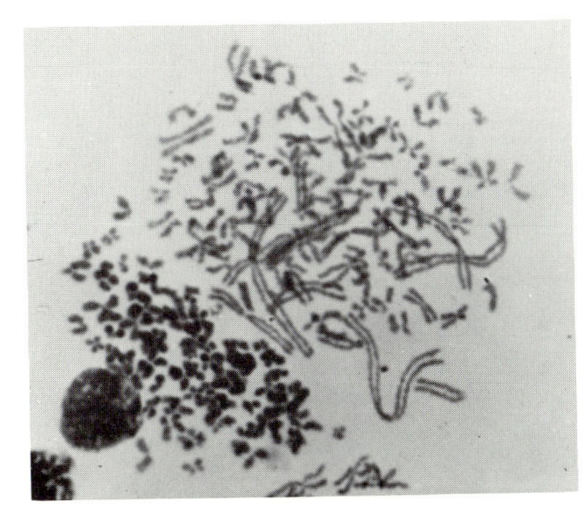

(c)

is. Various metabolic roles, such as messenger, regulatory, precursor, or structural functions, have been postulated for this rapidly turning-over nuclear RNA, but none has yet been unequivocally demonstrated. A specific kind of RNA, which has been found only in the nucleus, is the chromosomal RNA, which has short chain lengths of only 40 to 60 nucleotides and is bound covalently to chromosomal protein. The function of this RNA species also is still uncertain, although a gene regulatory action has been attributed to it (see page 158).

Transfer, messenger, and ribosomal RNAs leave the nucleus and exercise their metabolic functions in the cytoplasm. Table 6-3 lists some comparative statistics for the RNA of mammalian and bacterial cells.

The transfer RNAs of mammalian cells resemble those of *E. coli* (see Appendix A). They are similar in size and structure and are rich in methylated bases, although the spectrum of transfer RNAs in mam-

Table 6-3 Vital Statistics for RNA

	Mammalian Cell (Human Liver)	*E. coli*
Total amount (g/cell)	2×10^{-11}	10^{-13}
Total ribosomes per cell	6×10^{6}	30,000
Messenger-RNA half-life	Variable periods of 4-80 hours have been observed	1-3 minutes

malian cells may possibly contain a larger variety of forms. In addition, the patterns of methylation in the transfer RNAs have been demonstrated to be different in normal adult, as opposed to tumor and fetal, tissues, the latter having a higher proportion of methylated bases and containing certain new species of transfer RNA not found in normal cells of the adult. Recent experiments indicate that these differences are probably due to differences in the content of inhibitors of the methylating enzymes, which may differ in kind and amount in different tissues.

Difficulties have been encountered in following the activities of messenger RNA in mammalian cells. It is not yet known how this RNA, which is synthesized in the nucleus, is transported to the cytoplasm, finds its target ribosomes, and initiates protein synthesis. No pure messenger RNA has yet been isolated which will, of itself, stimulate initiation of specific protein synthesis in a cell-free system. Polysomes have been identified in mammalian cells, and RNA, which appears to have certain properties similar to those expected of messenger RNA, has been isolated from these structures. However, the extent of protein synthesis that can, as yet, be achieved by recombining the various components of protein-synthesizing systems extracted from mammalian cells, is far less than that obtainable from bacteria.

Measurement of the half-life of mammalian cell messenger RNAs has yielded a variety of species with a wide range of stabilities. Thus, messengers have been reported with half-lives of 4 hours in HeLa cells, 8 to 10 hours in erythroblasts, and up to 80 hours in adult rat liver. In *E. coli,* on the other hand, the messenger half-lives are more constant, and range around 1 to 3 minutes.

Mammalian ribosomal RNA has received intensive study. The genes for the ribosomal-RNA precursors are localized in specific chromosomal regions and are reiterated perhaps as many as 100 times. The RNA synthesized from these genes accumulates in the nucleolus, a round, densely staining structure that encircles the DNA regions involved. About 5 to 10% of the nucleolus is RNA, and there is also present a large amount of protein, which includes histones (see page 157) and RNA polymerase. In human cells, five chromosomes appear to be associated with the nucleolus, chromosomes 13, 14, 15, 21, and 22, which are the satellited chromosomes (see Chapter 2). Usually there is one nucleolus per haploid set of chromosomes, but fusion of nucleoli in the same nucleus can occur to reduce the number of these bodies.

The completed ribosome contains a large and a small subunit, with sedimentation constants of 60S and 40S, respectively. A 28S RNA particle is associated with the former, and an 18S RNA with the latter. The precursor of both these RNA species is a large 45S moiety, which is formed in the nucleolus and is processed by the following series of steps to form the ribosomal RNA entities:

$$\text{Nucleolar DNA + RNA polymerase} \longrightarrow \text{45S RNA}$$
$$\text{45S RNA} \longrightarrow \text{32S RNA + 18S RNA}$$
$$\text{32S RNA} \longrightarrow \text{28S RNA}$$

The large subunit, in addition, contains a 7S RNA, which also originates from the 32S intermediate, and a 5S RNA species which appears to arise independently. In the HeLa cell, at least, the genes for the 5S particle are located on chromosomes different from those which give rise to the 45S precursor. Not all of the RNA of the original 45S particle ends up in the ribosome, as shown by the fact that about 45% of the 45S particle has nonribosomal base sequences that are unusually high in guanine-cytosine base pairs, low in adenine, and possess few or no methyl groups. There is evidence suggesting that reiterated sequences may be involved.

When first formed, the 45S precursor RNA possesses only the four conventional purine and pyrimidine bases: adenine, uracil, guanine, and cytosine. Methylation of some of these bases rapidly occurs in the nucleolus. Methylation can also occur on particular ribose moieties, thus rendering those sites relatively insensitive to ribonuclease digestion. The methylation appears necessary for further processing of the 45S RNA, since, if it is prevented, no mature ribosomal RNA is formed.

The association of RNA and protein also occurs in the nucleolus. A variety of nucleoprotein particles, which appear to be precursors of the ultimate ribosomal subunits, are found in this structure. RNA transport from the nucleolus to the cytoplasm occurs rapidly. The various species of RNA are found in the cytoplasm within time periods after synthesis ranging from a few minutes to one hour.

Mammalian ribosomes (Table 6-4) are larger in size than those of *E. coli* and are different chemically, as evidenced by the larger percentage of proteins in their composition and by their resistance to inactivation by streptomycin. It is this selective specificity that makes streptomycin a useful antibiotic for treatment of bacterial infections of mammals.

Mammalian cells also contain ribosomes in the mitochondria. It is interesting to note in the data of Table 6-4 that, in some ways, mammalian mitochondrial ribosomes resemble those of *E. coli* more than they do those of the bulk of the mammalian cell cytoplasm. This fact lends support to the theory that mitochondria originated through a symbiotic combination of bacteria with higher cells.

Table 6-4 Comparison of Mammalian and Bacterial Ribosomes

Organism	*Ribosome Size*	*Inhibition of Protein Synthesis by*	
		Chloramphenicol	*Cycloheximide*
E. coli	70S	+	–
Mammalian cell	80S	–	+
Mitochondria	70S	+	–

Mention should be made of the recent demonstration that in mammalian cells infected with certain tumor viruses a reverse transcriptase enzyme is found which is capable of synthesizing DNA from an RNA template. This discovery would appear to have important implications for cancer and other aspects of mammalian cell dynamics.

The composition of DNA and its relationship to protein synthesis. Protein synthesis in mammalian cells provides many unsolved problems. In Table 6-5, the protein content and the rate of protein synthesis in a mammalian cell is compared with *E. coli.* Mammalian proteins are manufactured not only for carrying out specific functions within the cell but also for excretion into body fluids where they may serve functions at sites remote from their point of origin. Thus, some cells can manufacture extracellular enzymes, such as pepsin and trypsin; albumin for maintenance of osmotic pressure of the blood; antibacterial agents, such as lysozyme and antibodies; lubricating proteins, such as those which occur in the fluids bathing epithelial

Table 6-5 Vital Statistics for Protein

	Mammalian Cell (Human Liver)	*E. coli*
Total amount (g/cell)	5×10^{-10}	2×10^{-13}
Rate of synthesis (amino acids/sec)	4×10^{7}	1×10^{6}

surfaces; substances, such as collagen, which support body structures; and many more.

The mammalian cell has approximately 750 times as much DNA as *E. coli.* One might expect, then, that there might be about 750 times as many different mammalian proteins as those existing in *E. coli.* Nowhere near this number of mammalian proteins has yet been identified. The reasons for this may include the fact that many mammalian proteins may be present in cells in exceedingly small quantities, so that their isolation and identification is difficult. In addition, however, the possibility must be considered that appreciable amounts of the mammalian DNA may not code for protein but may have different functions, such as coding for regulatory RNA, serving as attachment sites for special molecules, or exerting a structural role in the chromosomal organization.

As discussed in Chapter 4, an appreciable fraction of mammalian DNA is reiterated. In the mouse, 10% of the total DNA is a highly repeated form, confined to the centromeric region of the chromosomes, whose unitary length is so short as to make it doubtful that it codes for protein. The existence of DNA reiteration of protein-producing genes would also result in fewer kinds of protein than would otherwise be expected.

It has been mentioned previously that a good protein synthetic system operating *in vitro* has yet to be prepared from mammalian cells. Apparently, additional and, as yet, unidentified factors are needed to cause messenger RNA to initiate protein syntheses on the ribosomes, even in the presence of fully active transfer-RNA molecules. It is conceivable that attempts to extract an active protein-synthetic system from mammalian cells are relatively ineffective because the extraction process activates lytic activities that destroy some needed component(s) of the system.

The histones are a particularly interesting group of proteins in the cells of higher organisms. They are completely absent in bacteria. They have a large complement of positively charged amino acids (25 mole percent of lysine and arginine) and are found in the nucleus specifically attached to the chromosomes. As recently as 1963, the number of different histones in cells was considered to be somewhere between 50 and 10,000. Newer studies indicate that the true number, at least in several cases, is probably 8.

Specific genes are selectively activated and inactivated as a cell progresses through its reproductive life cycle or as it assumes different differentiated states. By analogy with the action of the repressor molecule, which has been demonstrated to turn genes on and off in bacteria, one might expect proteins to be produced in mammalian cells that would selectively cover up or expose specific chromosomal regions in accordance with the cell's needs. Nuclei of higher organisms contain large quantities of histones, and, since these proteins are rich in basic amino acids, they can readily form complexes with DNA. It has been demonstrated that the addition of histones to DNA suppresses RNA synthesis and that the removal of these proteins from nuclei by means of trypsin can increase RNA synthesis. The structure of histones in cell nuclei undergoes modification by means of acetylation, methylation, and phosphorylation. The extent of such modifications can be demonstrated to change in specific ways when cells like lymphocytes are stimulated to divide. Similarly, an increase in histone acetylation occurs before the increase in RNA synthesis begins in regenerating liver cells.

Despite these indications that histones do combine with DNA so as to inhibit RNA synthesis, there is increasing evidence that these molecules may not be the primary gene regulators. Perhaps the most important reason for this view is the apparent lack of specificity in the structures of histones found in cells from different organisms and differentiation states. However, an interesting suggestion has recently been put forth. The short-chain-length chromosomal RNA is highly heterogeneous with respect to its base sequences and is bound covalently to chromosomal protein. The resulting complex is bound to histones which, in turn, bind to the DNA by ionic forces. These relationships have led to the proposal that the chromosomal RNA is involved in regulating gene action. Thus, when it attaches to a histone molecule, this RNA may provide the specificity needed by the protein in turning particular genes on and off.

Nuclei also contain significant amounts of acidic phosphoprotein, whose presence is highest in areas that are actively synthesizing RNA. These molecules may also operate in regulating the activity of chromosomes.

Other continuous processes. Mitochondria presumably reproduce only during late S and early G_2. Also, since mitochondrial structures include proteins derived from nuclear, as well as mitochondrial genes, there would appear to be regulatory interactions between the mitochondria and other cell structures, but these remain to be elucidated. As the cell proceeds along its reproductive cycle, its total mass increases steadily from G_1 until telophase, so presumably the membranes and structures related to the total cell volume are also being synthesized continually. Cells multiplying *in vitro* have been demonstrated to synthesize many other components, including steroids, polysaccharides, and a variety of conjugated proteins. The detailed study of these events in relationship to the life cycle promises to be a rewarding undertaking.

6-4 HUMAN IMPLICATIONS

This chapter has briefly outlined some of the new experimental and conceptual tools that are now being applied to an analysis of the molecular events involved in the reproductive cycle of mammalian cells. It is obvious that these studies are at an early stage. Nevertheless, the tools already available seem capable of producing valuable information. Understanding of the reproductive process will have important implications for processes such as those of embryonic development, postembryonic growth, cell reproduction in tissues like bone marrow and skin, which are constantly renewing themselves, wound-healing, and the action of drugs on these various functions. Cancer is an example of abnormal cell growth, and one can confidently expect that this subject will profit greatly from the coming developments in this field.

6-5 SUMMARY

Methods now exist for analysis of the biochemical events occurring at different points in the reproductive cycle of mammalian cells. From the study of the accumulation of cells blocked in various phases of the life cycle, one can determine the generation time of the culture, its degree of randomness, the length of time required for each of the four major divisions of the life cycle, and the points of action of various physical and chemical agents. Cells can be phased at different points in the cycle, and various biochemical events identified. Some processes, such as RNA and protein syntheses, occur continually throughout most of the life cycle. Others, such as DNA replication and synthesis of specific enzymes, occur for a limited period in specific phases. By fusing cells which have been phased in different periods, one can study the factors necessary for the development of certain phases of the reproductive cycle. Eventually it should be possible to map the sequence of necessary biochemical steps that convert a cell from the preceding state through successive states along the cycle.

While the general pathway DNA ⟶ RNA ⟶ protein is as vital in mammalian cells as in *E. coli*, more complexities exist in the former case: (1) mammalian DNA contains highly reiterated regions, many of which do not code for proteins; (2) any particular mammalian cell synthesizes only a small fraction of the total proteins coded for by the genome; (3) there is no counterpart in bacterial systems for the heterogeneous, rapidly turning-over nuclear RNA of mammalian cells; (4) bacteria do not contain histones, which are found attached to mammalian DNA.

Understanding of the molecular dynamics of the reproductive cycle in mammalian cells will have important implications for such human problems as embryonic development, cancer, and aging.

REFERENCES

Selected papers

Attardi, G., and F. Amaldi. Structure and synthesis of ribosomal RNA, *Ann. Rev. Biochem.* **39**, 183 (1970).

Baserga, R. Mitotic cycle of ascites tumor cells, *Arch. Path.* **75**, 156 (1963).

Beams, H. W., and S. Mueller. Effects of centrifugation on the interphase nucleus with special reference to the chromatin-nuclear envelope relation, abstract, ninth annual meeting, American Society for Cell Biology, November, 1969.

Bonner, J., M. E. Dahmus, D. Fambrough, R. C. Huang, K. Marushige, and D. Y. H. Tuan. The biology of isolated chromatin, *Science* **159**, 47 (1968).

Borek, E. In *Exploitable Molecular Mechanisms and Neoplasms,* Williams & Wilkins, Baltimore, Maryland, 1969, p. 163.

Darnell, J. E., Jr. Ribonucleic acids from animal cells. Bact. Rev. **32**, 262 (1968).

DuPraw, E. J. "DNA and Chromosomes." Holt, New York, 1970.

Fan, H., and S. Penman. Mitochondrial RNA synthesis during mitosis, *Science* **168**, 135 (1970).

Georgiev, G. P. The nature and biosynthesis of nuclear ribonucleic acids, *Progr. Nucleic Acid Res. Mol. Biol.* **6**, 259 (1967).

Howard, A., and S. R. Pelc. Synthesis of deoxyribonucleic acid in normal and irradiated cells and its relation to chromosome breakage, *Heredity* (suppl.) **6**, 261 (1953).

Huberman, J. A., and A. D. Riggs. On the mechanism of DNA replication in mammalian chromosomes, *J. Mol. Biol.* **32**, 327 (1968).

Johnson, R. T., and P. N. Rao. Mammalian cell fusion. II. Induction of premature chromosome condensation in interphase nuclei, *Nature* **226**, 717 (1970).

Kerr, S. J. Natural inhibitors of t-RNA methylases, *Biochemistry* **9**, 690 (1970).

Murphree, S., E. Stubblefield, and E. C. Moore. Synchronized mammalian cell cultures. III. Variation of ribonucleotide reductase activity during the replication cycle of Chinese hamster fibroblasts, *Exptl. Cell Res.* **58**, 118 (1969).

Penman, S., and C. Vesco. The cytoplasmic RNA of HeLa cells: New discrete species associated with mitochondria, *Proc. Natl. Acad. Sci.* **62**, 218 (1969).

Perry, R. P. The nucleolus and the synthesis of ribosomes, *Progr. Nucleic Acid Res. Mol. Biol.* **6**, 219 (1967).

Puck, T. T., and J. Steffen. Life cycle analysis of mammalian cells. I. A method for localizing metabolic events within the life cycle, and its application to the action of colcemide and sublethal doses of x-irradiation, *Biophys. J.* **3**, 37 (1963).

Rao, P. N., and R. T. Johnson. Mammalian cell fusion. I. Studies on the regulation of DNA synthesis and mitosis, *Nature* **225**, 159 (1970).

Smith, E. L., R. J. DeLange, and J. Bonner. Chemistry and biology of the histones, *Physiol. Rev.* **50**, 159 (1970).

Stubblefield, E., and S. Murphree. Synchronized mammalian cell cultures. II. Thymidine kinase activity in colcemide synchronized fibroblasts, *Exptl. Cell Res.* **48**, 652 (1969).

Terasima, T., and L. J. Tolmach. Growth and nucleic acid synthesis in synchronously dividing populations of HeLa cells, *Exptl. Cell Res.* **30**, 344 (1963).

Tjio, J. H., and G. Östergren. The chromosomes of primary mammary carcinomas in milk virus strains of the mouse, *Hereditas* **44**, 451 (1958).

CHAPTER 7

Epilogue: Differentiation and Behavioral Biology

This book has demonstrated how the development of *in vitro* techniques has already opened up large new areas in mammalian cell biology. The next immediate target of modern biomedical research is the mechanism of differentiation. Extensions of the techniques here described are already contributing important new understanding in this field.

The central problem of differentiation involves determination of the mechanism by which the expression of particular genes is controlled in mammalian cells. The cells of each tissue must develop step by step in their proper sequence and maintain activity of only those specific genes needed for proper functioning. The classic experiment of Gurdon demonstrated that even highly differentiated epithelial cells of the frog have preserved the genetic information necessary to produce the whole organism. Thus, when nuclei are transplanted from such highly differentiated cells into enucleated eggs, normal embryonic development occurs, leading to new adult animals capable of reproducing. Therefore, differentiation need not alter the genome in any permanent way, but, rather, simply regulate expression of the genes which are to be selectively active in the cells of the various tissues.

In vitro techniques utilizing molecular-biological approaches in a variety of systems are being applied to the analysis of this problem. For example, studies are in progress utilizing the lymphocyte, which is normally almost completely dormant biochemically. By treatment with appropriate agents (such as the plant glycoprotein, phytohemagglutinin, or a specific antigen to which the parent animal was previously immunized) this cell can be induced to adopt a high level of biochemical activity, including an increase in RNA synthesis, a wave of general protein synthesis, enlargement of the cell, reproduction, and the synthesis of gamma globulin.

Cells from several normal differentiated tissues can now be cultured *in vitro* under conditions in which the specific differentiated characteristics are retained, as, for example, the production of myosin by muscle cells and the synthesis of collagen by chondrocytes.

Sato and his coworkers have developed elegant systems in which the potentiality for biosynthesis of tissue-specific proteins can be maintained in cloned cancer tissue cultures for long periods. Cells from tumors which had retained some of the specific biosynthetic activities of their parental tissues were used for this purpose. Selective techniques were developed, in which the malignant cells were alternately grown in tissue culture and then in a susceptible animal. As a result of this sequence of growth experiences, it became possible to select cultures which demonstrated vigorous, long-lasting growth *in vitro* and would also continue to display a specific biosynthesis characteristic of particular differentiated tissues. Thus, it was possible to prepare cultures in which adrenocorticotropic hormone (ACTH) regularly induces synthesis of large amounts of steroid hormones.

Hormones have been found to be active inducers of a variety of enzymes in mammalian cells. Tomkins and his colleagues have demonstrated that the cells of a rat hepatoma culture, which normally contains small amounts of the enzyme tyrosine transaminase, can be induced by the steroid dexamethasone, *in vitro,* to increase the synthesis of this enzyme by a factor of 15-fold. Life-cycle analysis of these cells demonstrated that the enzyme can be induced only in the last two-thirds of the G_1 phase and in the S phase of the cellular life cycle. Study of the kinetics of this induction led Tomkins to postulate that an important regulatory step in this system is a degradation of the otherwise stable messenger RNA after its combination with a specific regulatory protein.

A number of studies demonstrating various kinds of cell-cell interaction have also recently appeared. For example, Sachs and his coworkers have shown that mature macrophages and granulocytes produce specific substances that inhibit the development of their precursor cells, so exercising a feedback control of the cells in these differentiation pathways. Other investigators have demonstrated direct effects of cell-to-cell contact in experiments utilizing nutritionally deficient cell mutants. Thus, a mixed culture was prepared of a mutant lacking inosinic phosphorylase activity (IPP^-), which could not incorporate tritium-hypoxanthine into nucleic acid, and of a normal cell possessing this enzyme (IPP^+). It was found that the deficient mutant would incorporate the labeled hypoxanthine into nucleic acid if it were allowed to establish cell-to-cell contact with an IPP^+ cell.

Various techniques have been introduced by which the existence of specific recognition elements on the cell surface can be established. For example, Moscona has shown that dispersed cells from such embryonic tissues as liver, kidney, and neural retina will spontaneously and specifically reaggregate to form structures highly characteristic of the tissue of origin.

Cancer represents a distortion of normal differentiation processes. Here cell multiplication does not confine itself to the needs of the particular tissue but continues without limit, piling up more and more cells that cannot be eliminated, so that eventually disorganization of body structures ensues. At present, it seems likely that the change that causes a normal cell to become cancerous could well be a mutation either at the level of individual genes or of the chromosomes. Several carcinogens, such as the nitroso compounds, have been shown, by means of the techniques described in Chapter 3, to be highly mutagenic at both levels for mammalian cells. One of the most interesting classes of carcinogens is that of certain mammalian viruses which are able to produce a variety of different kinds of cancer by a process called transformation. Indeed some workers have postulated that all cancer may be due to virus action. Proof or disproof of the universality of this hypothesis is difficult to demonstrate. However, the variety of actions of viruses in mammalian cells makes it possible to reconcile viral carcinogenesis with a more general mutational hypothesis. Viruses can incorporate their genetic information into the cell genome. Under these conditions, the genetic constitution of the cell is altered, and delicate regulatory mechanisms, which might be necessary to keep cells from behaving malignantly, may well be upset. In addition, many viruses have been demonstrated to break chromosomes of the host cells and to introduce chromosomal aberrations like those known to be the cause of specific malignancies. Finally, as we have seen in Chapter 3, viruses can cause cell fusion, in which case the tetraploid cell formed may readily lose chromosomes so as to become aneuploid and therefore possibly cancerous.

One of the most intriguing new leads in the study of cancer involves the phenomenon of contact inhibition, which is usually lost by a cell when it becomes malignant (see Chapter 1). The existence of methods for altering a cell's ability to grow in the contact-inhibited state by use of specific chemical agents, as well as by treatment with viruses, makes it seem hopeful that important illumination in the molecular biology of this process and its relationship to malignancy may soon be forthcoming.

Other studies offering promise in unraveling aspects of normal and pathological phenomena connected with differentiation include (1) study of specific interaction between various RNAs and proteins with different parts of the genome in cells being induced to form new enzymes connected with differentiation behavior, (2) the use of virus-mediated cell fusion to determine how differentiation properties will be affected by hybridization between carefully chosen cell pairs, (3) analysis of the function of reiterated DNAs in differentiation processes, (4) study of the roles of different kinds of transfer RNAs in differentiation processes, (5) study of factors capable of affecting the rates of transcription and translation of various regions of the genome, and (6) molecular study of the fate of hormones and other inducing agents when they make contact with target cells.

The answers to the new experimental questions, which these techniques make possible, hold enormous promise for human health. One can reasonably expect that the advances to be achieved will contribute understanding to the classical problems of aging, cancer, and tissue and organ healing and regeneration. It is possible that approaches of this kind may ultimately prove to be far more practical, in many kinds of human surgical processes, than organ transplantation.

One would hope that it may be possible to intervene in problems of abnormal human development so as to bypass the deleterious effects of gene mutations or chromosomal aberrations. Even more exciting is the possibility of achieving control over some of the normal developmental processes so as to produce human beings with increased competence in certain ways, as, for example, in the ability to produce antibodies to invading agents.

With developments in human genetics already under way, and the great promise for understanding differentiation processes which seems imminent, man will soon possess tremendous new biolcgical powers. In the coming era, we shall see a large measure of conscious control of evolutionary and developmental processes in man and other living forms. It will be perhaps the greatest challenge that man has yet had to face, to use these powers to increase the potentialities for human fulfillment.

There are indications on the horizon that growth of modern biology may afford man some means for better understanding of his own complex nature, and perhaps, therefore, of controlling his new powers more wisely. New developments in the field of behavioral biology, and the possibility that the newly emerging molecular biology of mammals may accelerate these developments, raises the possibility for new insights into the nature of human drives and emotional needs. Only a few examples will be cited here.

The study of population interactions in many animal forms, from beetles to mammals, is providing new understanding of behavior patterns due to competition and aggressiveness that raises the most soul-searching questions to students of human society. Innate behavioral patterns that respond to triggering stimuli have been found in many animal species: (1) The imprinting phenomenon, whereby newly hatched ducklings are induced to recognize and accept as a mother any figure that is appropriately presented to them, (2) the patterns of territorial defense in birds and other animals, and (3) the development of competition patterns in animal societies presented with specific environmental challenges are representative areas that indicate the complexity of the interactions between innate genetic instruction and social and environmental experience.

Certain of these behavioral patterns have been demonstrable in every animal species studied. There is little question, therefore, that similar drives also exist in man. However, man appears to be unique among the animals in requiring a system of values to regulate his basic biological drives and to build them into a fulfilling life pattern. Moreover, in man, the enormous capability for learning introduces a flexibility that is different in kind from that of any other animal species. The extent to which expression of the genetic behavioral patterns in man is controlled by developmental experience is one of the most fundamental questions whose answers hold the deepest implications for the future.

Some of the behavioral effects of high population density on small mammals have been shown to be accompanied by changes in hormonal levels of the adrenals and the gonads, and at least some of the crowding effects on behavioral patterns can be minimized by administration of specific chemical agents. Careful studies both in man and in the nonhuman primates have demonstrated the need for certain kinds of social contact in young animals for normal development to take place. Thus, human babies, as well as young monkeys, who are deprived of such social interaction at an early age develop clear pathological symptoms. There is a suggestion from studies in rats that responses to stress of the adrenal gland may be strongly affected in a lasting way by the experiences of infancy. Studies in mice have demonstrated that nutritional deprivation early in development can have irreversible, deleterious effects on the learning ability of the animals in later life. While these fields of research are still young, there is reason to hope that a fruitful fusion of the biological and social sciences is under way.

Synthesis between many areas of the new molecular biology and the mechanism of psychological processes also appears to be on the horizon. Cell-culture techniques like those described in earlier chapters are being employed to explore gene-expression mechanisms and the role of cell-cell interaction in neural development. By appropriate experimental manipulations, it is now possible to elicit biochemical and electrical properties characteristic of differentiated neurons, from cells of nervous system tumors grown in culture. The well-known availability of large numbers of new drugs, such as the hallucinogens, that exercise specific actions on nervous-system functions, and the concomitant studies now in progress to delineate their biochemical action in cells, raise the hope that it will be possible to isolate specific biochemical steps responsible for particular mental operations. One of the exciting developments in this field has been the recent demonstration that mammals may possess two different kinds of memory systems with different time characteristics. One is a short-term memory, which decays rapidly. Normally, information can be transferred from the short-term to the long-term memory, where it will remain stable for considerable periods. Experiments have been carried out which appear to imply that transfer of information from the short-term to the long-term memory can be inhibited by some (though not by all) agents that prevent protein synthesis. The molecular basis of memory and other nervous-system functions appears to be one of the most adventuresome fields of the new biology.

While animal behavior, in general, and human behavior, in particular, still represent an exceedingly difficult area to study scientifically, the indications are already available that important illumination in these areas will soon be possible. The history of all science has shown a repeating pattern of development, in which a field initially vague, mystical, and accessible only to flashes of inspiration and intuition, gradually becomes scientifically operational with clear, hard facts and powerful means of accomplishing particular ends. During the Greek period, the constitution of matter was a vague and mystical region for speculation by poets, philosophers, and religionists. The growth of physics has been one of the most exciting experiences of the human mind. In our time, we have seen biology become transformed from a vague, descriptive discipline into a powerful experimental and conceptual science. There would seem then to be every reason to believe that both psychology and sociology can take the same path when the necessary groundwork has been prepared.

With the growing understanding of the nature of his genetic constitution, man is acquiring great powers over his own future evolutionary progress. This situation is frightening in view of the limited understanding that he now seems able to bring to bear on problems of far less complexity. Acquisition by man of these tremendous new powers over his own condition makes still more acute the question that first arose when the release of nuclear energy became possible. Will mankind be able to acquire sufficient wisdom and practical knowledge to harness these powers for human betterment, or will the rigidities of human social structures and behavioral patterns force the coming of a holocaust on a planetary scale?

The rapidity with which scientific advances are affecting human life on a large scale makes untenable continuation of a policy of irresponsibility of the scientist for the social consequences of new discovery. No one else is in a position to advise society and, particularly, its lawmakers and administrators of the short- and long-term implications of scientific advances for man. No one can, as well, explain the need for extension of operational principles into large areas of human decision-making, at least to the extent of forcing recognition of the degree of uncertainty inherent in the parameters by which the decisions are determined.

Throughout most of the evolutionary history of an animal like man, it is easy to see how the existence of emotionally triggered reactions and automatic behavioral patterns afforded an enormous advantage. However, these same biological mechanisms, which once gave our ancestors a tremendous advantage over their competitors in the jungle, now pose a threat to a human way of life in our highly complex industrial civilization.

Man's all-too-rapid scientific and technologic development has caused some of his innate as well as his culturally habituated regulatory processes to become obsolete and to pose new threats to his well-being and possibly even to his continued existence. We are now engaged in a race to determine whether man will be able to develop, by means outside of biological evolution, the necessary changes in his response systems that will make continued human life possible in a world in which unbearable population pressures, extreme economic disparities, the threat of irreversible pollution, and the possibilities of nuclear and biological warfare have become imminent realities.

Difficult as are the acquisitions of the practical tools for achieving greater flexibility in human responses to social stimuli, the problem of developing the wisdom to use these and other scientific advances effectively appears even more difficult. The scientific-technological revolution during the last 300 years has promoted a divorce of science and philosophy, and disintegration of many of the classical human value systems. The great intellectual expression of our age is found in science. However, through specialization and the lack of communication, the sciences themselves have tended to become fragmented and isolated. The one universal aspect common to both science and philosophy, namely, the critical testing of the degree of truth of any proposition and the evaluation of the uncertainty of conclusions drawn from it, is largely ignored in most aspects of life, and particularly in decisions affecting public policy. As a result, prediction in human social affairs is primitive, and decision-making in society all too often is based on traditional principles which have lost much of their meaning. Moral values, such as the brotherhood of man, usually remain expressed in terms of traditional supernatural frameworks, many of which are mutually exclusive, and at least some of which are expressed in terms antagonistic to the rationalistic spirit of the time. Material values alone are strengthened, and people are left confused and neurotically predisposed by the resulting distortion of human guidance systems.

Man as a biological entity differs from all other living things in his dependence on long-term goals that command his greatest energies. These need to be reexamined and reformulated afresh in our time and to be translated into means for accomplishment, compatible with the spirit and opportunities of this unprecedented new era. It would appear that the only healthy biological solution to the dilemma that confronts modern man is a new intellectual and moral synthesis, giving expression to human values in a way that is consistent with the needs of man as a biological and a social organism, and making available to all mankind the almost limitless potential for development which the new biological science appears to offer.

REFERENCES

General

Dulbecco, R. Viruses in carcinogenesis, *Ann. Intern. Med.* **70**, 1019 (1969).

Gurdon, J. B. Transplanted nuclei and cell differentiation, *Sci. Amer.* (December), 24 (1968).

Hess, Eckhard. Imprinting in animals, *Sci. Amer.* (March), 81 (1958).

Kohn, A., and P. Fuchs. Communication in cell communities, *Current Topics Microbiol. Immunol.* **52**, 94 (1970).

Lorenz, Konrad. The evolution of behavior, *Sci. Amer.* (December), 67 (1958).

Lorenz, Konrad. "On Aggression," Harcourt, Brace & World, New York, 1966.

Park, Thomas. Beetles, competition and populations, *Science,* **138**, 1369 (1962).

Platt, John R. "The Step to Man," Wiley, New York, 1966.

Ursprung, H. (editor). "The Stability of the Differentiated State," Springer-Verlag, New York, 1968.

Weiss, P. A. Dynamics of development: experiments and inferences, in "Selected Papers on Developmental Biology," Academic, New York, 1968.

Williams, E. D. Tumors, hormones, and cellular differentiation, *Lancet* 1108 (1969).

Selected papers

Brown, D. D., and I. B. Dawid. Developmental genetics, *Ann. Rev. Genetics* 3, 127 (1969).

Green, M. Oncogenic viruses, *Ann. Rev. Biochem.* **39**, 701 (1970).

Gurdon, J. B., and H. R. Woodland. The cytoplasmic control of nuclear activity in animal development, *Biol. Rev.* **43**, 233 (1968).

Konigsberg, I. R. Clonal analysis of myogenesis, *Science* **140**, 1273 (1963).

Moscona, A. A. Analysis of cell recombinations in experimental synthesis of tissues *in vitro, J. Cell. Comp. Physiol.* **60** (suppl. 1), 65 (1962).

Paran, M., and L. Sachs. The continued requirement for inducer for the development of macrophage and granulocyte colonies, *J. Cell Physiol.* **72**, 247 (1963).

Rutter, W. J., J. D. Kemp, W. S. Bradshaw, W. R. Clark, R. A. Ronzio, and T. G. Sanders. Regulation of specific protein synthesis in cytodifferentiation, *J. Cell. Physiol.* **72** (suppl. 1), 1 (1968).

Sachs, L. Feedback control of the development of hematopoietic cell clones *in vitro,* and the mechanism of leukemogenesis, *Can. Cancer Conf.,* Proceedings of the Eighth Canadian Cancer Conference, Honey Harbour, Ontario, 1968.

Sato, G., and V. Buonassisi. Hormone-secreting cultures of endocrine tumor origin, *Natl. Cancer Inst. Monograph* **13**, 81 (1964).

Subak-Sharpe, H., R. R. Burk, and J. D. Pitts. Metabolic co-operation between biochemically marked mammalian cells in tissue culture, *J. Cell Sci.* **4**, 353 (1969).

Tomkins, G. M., T. D. Gelehrter, D. Granner, D. Martin, Jr., H. H. Samuels, and E. B. Thompson. Control of specific gene expression in higher organisms, *Science* **166**, 1474 (1969).

APPENDIX I A REVIEW OF CHEMICAL REACTIVITY AND THE MOLECULAR BIOLOGY OF THE SIMPLEST LIVING CELLS

The chemistry of living organisms appears on first examination to be totally different from that of nonliving systems. Yet both obey the same laws of physics and chemistry. It is worthwhile, then, to review the basic features of chemical reactivity in order to make clear the basis for the chemical versatility of living forms.

All molecules are made up of atoms which contain a central, positively charged nucleus, which constitutes more than 99.9% of the atom's mass. Around the nucleus, distributed in shells, are exceedingly light, negatively charged electrons, present in a number such as to exactly balance the positive charge of the nucleus. The diameter of the nucleus is approximately 10^{-13} cm, while that of the outermost electronic shell is about 10^{-8} cm, or 1 Å. Therefore, the volume of an atom, for the most part, consists of empty space, in which powerful forces operate to maintain its dynamic structure. The ordinary incompressibility of solid and liquid matter is not due to the fact that the masses of the atomic parts actually come in contact, but rather to the repulsive effect of the electric fields from the planetary electrons that resist attempts to bring atoms close together.

Atoms can combine with each other to produce molecules. Certain basic rules of atomic combination exist which assign to each atom

Figure I-1 Representations of typical small molecules important in living processes. The shaded regions indicate the approximate extent of the atomic radii, which, for the atoms most frequently found in biological systems, range from 1.2 to 1.9 Å. In the glycine molecule, the two hydrogens depicted as touching the carbon and nitrogen atoms, respectively, actually lie outside the plane of the paper.

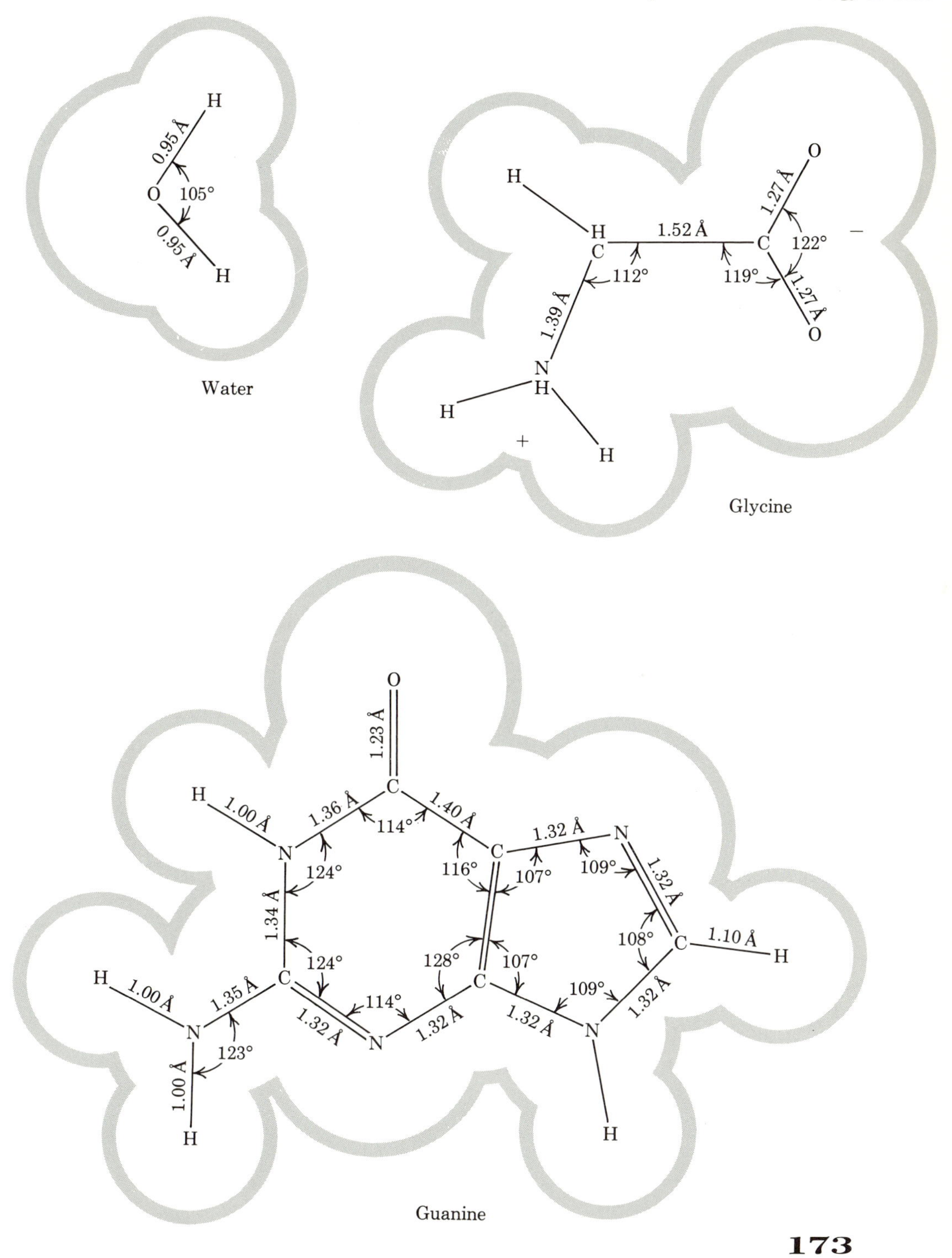
H
0.95 Å
O
105°
0.95 Å
H
Water
H
H
C
1.52 Å
C
O
1.27 Å
122°
–
112°
119°
1.27 Å
O
1.39 Å
N
H
H
H
+
Glycine
O
1.23 Å
C
H
1.00 Å
1.36 Å
114°
1.40 Å
N
C
1.32 Å
N
124°
116°
107°
109°
1.32 Å
1.34 Å
108°
C
1.10 Å
H
124°
128°
107°
H
1.00 Å
1.35 Å
C
C
109°
114°
1.32 Å
1.32 Å
1.32 Å
1.32 Å
N
N
N
123°
1.00 Å
H
H
Guanine

a valence or combining affinity for other atoms. The distance between two bonded atoms, the energy needed to break such a bond, and the angle between any two bonds formed simultaneously by a single atom are fixed; and, while bond distortion from the optimal position can occur, the process requires energy input. Therefore, the architecture of any molecule is a well-defined concept. Typical molecular dimensions are shown in Figure I-1.

Except for the rare gases, such as helium and neon, free atoms are almost never found in nature, because they tend to form stable chemical bonds with other atoms and so become incorporated into molecular aggregates. The ordinary molecules of daily experience, such as those of water, sugar, alcohol, or paper, consist of specific atomic configurations that are stable in air at room temperature, but this stability does not always mean that the molecules in question are necessarily at the lowest energy state that can be achieved by the given number and kind of atoms they contain. Of the examples mentioned, water is indeed in such a state. However, while paper and sugar appear to be stable, both of these, if properly stimulated, can unite with the oxygen of the air to undergo a combustion that releases an enormous amount of energy. As the end result of such a reaction, the atoms of the paper and of the oxygen are rearranged to from new molecules, carbon dioxide and water, which do represent the state of lowest energy or highest stability that can be achieved for the atoms of carbon, hydrogen, and oxygen.

That molecules which are potentially unstable can nevertheless remain unchanged for indefinitely long periods at room temperature, is of central importance in both the living and nonliving chemical economy of the earth. Were it not for this fact, all the chemical reactions on earth that could occur would occur spontaneously, and the entire earth thereafter would be lifeless and chemically inert. If a stable bond is formed between two atoms, it must be broken (or at least weakened) before the atoms will be free to reestablish new bonds, even though these may be more stable still. The energies needed to break ordinary chemical bonds (i.e., covalent bonds) lie in the range of 50 kilocalories per mole. At room temperature, virtually no molecules have enough energy to break such a bond spontaneously. Therefore, a large initial energy investment is required. The existence of this energy requirement, which is called activation energy, is the reason why gasoline or paper must first be ignited before it will

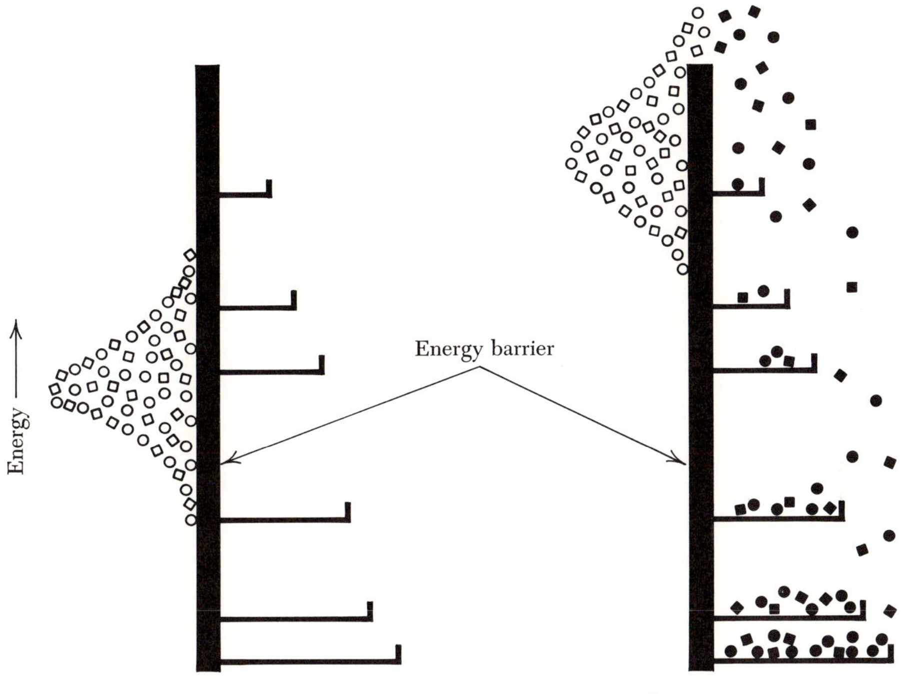

Figure I-2 Schematic representation of how a chemical reaction that liberates energy still needs an initial heat input (activation energy) in order to start. Unless the molecules are initially raised in energy to the top of the barrier, they cannot fall down to the final energy states even though these are lower than the initial one. As the molecules fall down from the barriers, the various energy levels, each representing a different molecular state of the component atoms, become filled. Other things being equal, the lower the energy level of any state, the greater the percentage of the products which end up in it. (Redrawn from T. T. Puck, The biological sciences, in "The Great Ideas Today," Encyclopedia Britannica, Inc., Chicago, 1967, with permission.)

burn in air. The requirement for this initial heat input to overcome the energy of activation is illustrated in Figure I-2. When a match is applied to paper or gasoline, the molecules are heated to a point where the random energy of their thermal motions suffices to break or loosen the chemical bonds in an appreciable proportion of these molecules. These activated molecules then react with the atmospheric oxygen, and the resultant combustion liberates much more energy than the initial input. This liberated energy then raises the temperature of additional molecules of the fuel, which in turn become activated and react, and so prepare still another fraction of the fuel for combustion. This cascade of activation ultimately brings about the combustion of all of the fuel. In the case of an explosion, the entire process is consummated in a small fraction of a second. If a reaction must be carried out which never produces heat, but rather must absorb energy in order to proceed (i.e., an endothermic reaction), the same considerations apply except that the external energy must be supplied continually, instead of just at the beginning of the reaction.

From the foregoing discussion, it might be erroneously concluded that the only quantity that determines the direction a chemical reaction shall take is the total energy released. Actually, however, there is another parameter which is equally important, that is, the gain in randomness which accompanies any reaction. Any possible reaction can go either in the forward or the reverse direction. Energy considerations tend to drive the reaction in the direction that is accompanied by the greatest release of stored energy. However, in addition, the direction taken by the system is also strongly influenced by the relative ease with which the interacting atoms can find each other. In the example shown in Figure I-3, we consider the spontaneous dissociation of the molecule AB into free A and B atoms. Dissociation of each molecule of AB requires only that the energy considerations be favorable. On the other hand, association of the atoms A and B into a new molecule cannot occur, regardless of the energy considerations, until an atom of A and one of B find each other under conditions of proper orientation. We say that a set of AB molecules has a greater information content than a mixture of A and B atoms, since, in the former case, specifying the location of any A atom automatically specifies that of its accompanying B. In the random mixture of A and B atoms, however, specifying the location of any A atom tells nothing about that of any B atom. The effect of this probabilistic feature is to tend to drive reactions in the direction of the greatest information loss (or randomness gain). The actual driving force of any reaction, then, is made up of two contributions, that of the energy released and the information lost.

We can now understand the characteristics of chemical reactions as they usually occur in the test tube. At room temperature, most chemical reactions will not proceed appreciably. (This does not apply to ionic double-decomposition reactions, since, in this case, no new

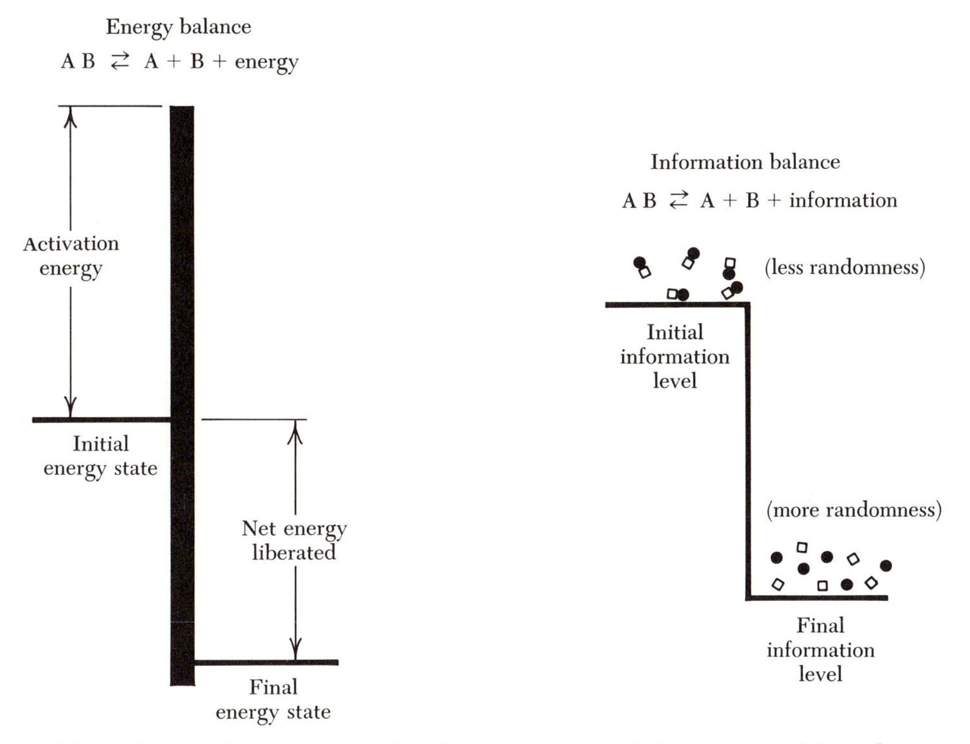

Figure I-3 *Demonstration that the combination of the energy and the information balance determines the direction in which chemical reactions go. Reactions will proceed spontaneously in the direction yielding the largest net liberation of energy and the largest decrease in information (or increase in randomness). While the final state attained depends only on the net change in these two quantities, an initial input of activation energy may be necessary to make the velocity of the reaction appreciably large. Increases in temperature make more energy states accessible, including those leading to dissociation and recombinations. (Redrawn from T. T. Puck, The biological sciences, in "The Great Ideas Today," Encyclopedia Britannica, Inc., Chicago, 1967, with permission.)*

chemical bonds are established, but rather the electrically charged ions simply rearrange their partners.) In order to cause a chemical reaction to proceed with appreciable velocity, the temperature must be raised. This increases the random kinetic energy of all the molecules and so increases the energy of the molecular collisions. Hence, chemical bonds throughout the molecule are activated randomly, and a number of different possible reactions will occur, as represented by the filling up of the intermediate energy levels in the diagram of

Figure I-2. By and large, the states representing products lowest on the energy scale and lowest on the information scale will tend to be most highly populated, since molecules which end up in these states will be less readily raised again to the top of the potential barrier. However, there will be a certain probability for filling all of the intermediate states shown in the diagram, so that even if one begins the reaction with the purest possible reactants, a variety of end products is usually obtained. The chemist then uses his sophisticated arts of purification to isolate, from the resulting mixture, the particular product at which he has aimed.

While some chemical reactions which occur in the test tube may behave in a somewhat more orderly fashion than that described here, the great majority of reactions that involve changes in chemical bonding do conform to the picture presented. The reason for this is that the reaction is carried out entirely through blind, random molecular collisions which make it usually difficult and often impossible to discriminate between desirable and undesirable reactions of the molecules involved. Finally, reactions carried out in the test tube usually begin with maximal velocity, since the concentrations of the reactants are maximal at the beginning. However, as time goes on, the concentrations of the reactants fall and those of the products, which can reverse the process, gradually rise. Therefore, the net reaction velocity will tend to fall steadily with time after the reactants have been mixed.

Often the chemist can expedite reactions by the use of a catalyst which lowers the activation energy required and so reduces the temperature needed for bond activation. However, the number of effective catalysts available to the organic chemist is sharply limited, their ability to lower activation energies is highly variable, and they are often susceptible to poisoning processes. Finally, catalytic action of a molecule often makes it able to participate in a variety of different reactions so that it is not easy to obtain 100% conversion of a given molecular species into a single desired form.

Since thermal energy is transmitted randomly from molecule to molecule, a complex mixture of reaction products is usually obtained in the course of the blind collisions that occur in a reaction system where heat is used to overcome the activation energy. In the course of the random molecular encounters, any bond in any molecule of a mixture may become free to react, and many of the new bonds which form will again be broken. It may also be noted that only a small fraction of the theoretically maximum quantity of useful work can be realized from the energy released by such a reaction. A large fraction of this released energy is lost as heat to the surroundings, and only if very elaborate apparatus is available, such as a turbine driven by steam at extremely high temperatures, can a large fraction of the theoretical work energy be extracted from such a process. Engineering equipment of this kind, though expensive and elaborate, is nevertheless often employed, as in the burning of fuel for the commercial production of electricity. It should be noted, however, that at high temperatures, energy-liberating reactions seldom go as far toward completion as they could go at a lower temperature. Therefore, in addition to all their other disadvantages, nonliving processes for conversion of chemical energy into work are usually relatively inefficient.

Physicists and chemists have long recognized the relatively inefficient nature of the process by which most chemical reactions are carried out in the test tube and the industrial retort. However, there seemed to be no way by which the inefficiency of such disorderly processes could be replaced by a more orderly arrangement. It was commonly pointed out that the fundamental uncertainties inherent in the structure of matter itself limit any attempt to bring about reactions between molecules in a more orderly fashion. Thus, the randomness of thermal motion itself confers a basic disorder on any ensemble of molecules. In addition, the uncertainty principle of quantum mechanics defines another basic randomness connected with the position and velocity of atoms and their constituent electrons whose rearrangements underlie all chemical reactions. It remained for the development of understanding of the operation of living systems to illustrate how, despite these fundamental uncertainties inherent in the atoms and molecules themselves, it is possible to marshal atoms and molecules so as to carry out chemical reactions in a highly orderly fashion.

The chemistry of living processes. Biological systems contain the same atoms as the nonliving materials and often carry on the very same chemical reactions, such as fuel combustion, that occur in nonliving systems. Yet the chemical reactions occurring in living cells display characteristics fundamentally different in kind from those that occur randomly in the test tube.

The energy for all of the diverse chemical activities of the cell comes from the combustion of energy-rich molecules, such as carbohydrates. In a fraction of a second, the energy liberated from such fuel molecules may be so large that were it applied in random fashion, it could cause extensive bond breakage and molecular disorganization of the very substance of the cell itself. The process may be compared to the burning of coal in a locomotive which is itself built of wood. Yet, far from destroying the molecules which constitute the cell structure, the energy so liberated is smoothly and effectively harnessed to carrying out the highly diverse activities of the living processes. The rates of the chemical reactions which occur inside the cell achieve extremely high velocity, yet all the reactions occur at the low temperature of the body. Moreover, despite the molecular heterogeneity inside the cell, each chemical transformation is accomplished with a complete absence of side reactions and without interference by the many different molecular species present in the same milieu. Thus, each chemical reaction produces one and only one set of products even though many of the reactions release enormous amounts of energy. Further, when such energy-liberating reactions occur in living cells, a fairly large fraction of the maximal yield of useful work is often obtained. Finally, chemical reactions inside cells are carried out under the aegis of a system of powerful, automatic feedback controls that determine the nature of the reactions that take place at any time and that modulate the reaction velocity at each moment to conform to the needs of the organism. Compared to the sophisticated mode of molecular manipulation that is achieved in even the simplest living cell, modern nonliving chemistry would appear to be primitive indeed.

The power of biological systems to direct molecular transformation in highly specific ways is evident in the process of biological reproduction. The degree of stability inherent in the molecular replicative system of living cells and viruses exceeds by many orders of magnitude that of any man-made copying systems, regardless of whether the patterns to be copied are formed by light or sound, or by other molecules. The action of such systems is illustrated by the life cycle of a virus. T2 bacteriophage, a virus that attacks certain bacteria, is a structure of approximately 650 Å in diameter and 2000 Å long, containing nucleic acid and proteins. The large protein headpiece forms a container for the nucleic acid and terminates in a tail-like structure, consisting of a hollow protein tube ending in a plate with thin leglike extensions. The structure contains many millions of atoms linked to each other by chemical bonds in a completely orderly and specified fashion. When placed in a salt solution, the virus particles are inert and are buffeted around in random Brownian motion, exactly like any nonliving particle of the same size. However, if bacterial host cells are now added to this solution, an exceedingly complex set of events is set in motion. At first, the virus particles continue to move randomly as before, except that from time to time they collide with a host cell. If a virus collides with an immune cell, it bounces off without change. But if it collides with a cell capable of acting as a host, bonds are immediately established between molecules on the two surfaces. Within 1 minute, a hole forms in the wall of the cell, and the nucleic acid contained in the head of the virus is injected through the hollow tail into the body of the cell. Then, with almost equal swiftness, the hole in the cell wall is resealed. The empty protein capsule of the virus remains functionless on the outside of the cell. The injected nucleic acid now halts the synthesis of bacterial components, and the virus begins to fashion molecular machinery for its own reproduction. Finally, the viral nucleic acid and protein are faithfully copied and assembled to form new whole virus particles inside the cell. Soon thereafter, the entire bacterial cell wall disintegrates and approximately a thousand new viruses, each one indistinguishable from the original infecting particle, are poured out into the surrounding solution. The entire process, from the initial collision between virus and host cell to the final discharge of the newly replicated viruses, requires only 22 minutes. In this short time, billions of new specific chemical bonds have been established from component atoms taken from much simpler molecules. An assembly process, as complex and accurate as this one, would be remarkable even if the individual components were several centimeters in length, that is, of a size that is readily manipulable. The fact that it takes place on the smallest possible scale for stable matter, so that the operative units consist of individual atoms and molecules, makes this behavior phenomenal. Obviously, highly directed cellular chemistry requires a much more orderly kind of interaction than the random processes that govern chemical reactions in the test tube.

The manner in which living systems have learned to reduce the randomness of molecular processes involves exploitation of atomic forces much weaker than those of the ordinary chemical bonds. These interactions, which involve energies between 0.5 and 5 kilocalories per mole, are of several kinds, but all involve attraction between regions in which nonuniform distributions of electrical charge are established and result in small but appreciable net attractive forces. In contrast to most chemical bonds which cannot be broken by thermal energy at room temperature, these weaker interactions are rarely able to hold molecules together against thermal collisions, unless the temperature is lowered or unless several such weak bonds cooperate in maintaining a specific molecular association. One example of such interaction is the so-called dipolar bond, in which the centers of positive and negative charge in a molecule do not coincide. The molecule itself forms a dipole, and an attractive force exists between this molecule and others in which a separation of charge also exists. Another case occurs when two atoms that have no permanent dipole approach sufficiently close together. As a result of their mutually fluctuating charge distributions, the atoms polarize each other to form induced dipoles, which then exert the net attractive forces that underlie the van der Waals bond.

Other examples are: (1) hydrogen bonding, in which a somewhat positive hydrogen atom forms a bonding bridge between two electronegative atoms in the same or different molecules; (2) charge-transfer interaction, in which an electron is partially or completely donated by one complex molecule to another, where it can assume a lower energy level than was available in the donor molecule, with the result that the two entities are now held together by electrostatic attraction; and (3) hydrophobic interactions that are engendered when a molecule, such as a protein, with many nonpolar groups, is dissolved in a polar substance, such as water, in which case the aromatic residues tend to associate with each other and become buried in the interior of the molecule.

The principal features of the weak bonds which are important for the present purposes are as follows:

(1) They are effective only at atomic distances. Thus, in these weak interactions, the forces engendered fall off extremely rapidly with distance so that, unless the atoms are virtually touching, the magnitude of the forces becomes very small indeed.
(2) They involve little or no activation energy. Bond formation and dissolution may therefore occur very rapidly at low temperatures in response to changes in the molecular and ionic environment.
(3) They may be simple or complex. The simple interactions, involving attraction between a single positively charged site of one molecule and a negatively charged one of another, allow only a limited specificity. On the other hand, when several sites on the two molecules are involved, they can form a specific geometrical pattern that may include a variety of different kinds of physical interactions. This permits highly selective molecular attachment processes.

(4) They determine solubility in water and other solvents. Molecules with ionic charges or large dipoles are readily soluble in the highly polar substance, water, while hydrocarbons are soluble in nonpolar solvents. This rule is only true, of course, if the solute and solvent molecules, respectively, do not tend more to associate with themselves than with each other. The solubility of molecules with both hydrophilic and hydrophobic groups is more complex. Thus, a single hydrophilic group, such as a hydroxyl or an amino group, suffices to make a molecule with as many as five or six carbon atoms water soluble.
(5) To a large extent, they determine molecular shapes and orientations. Thus, in aqueous solution, a molecule with many polar $—NH_2$ or —OH groups will adopt a configuration in which these groups are turned outward to interact with water molecules or with other polar groups with which they can form complementary bonding structures. These other groups may be in the same or in different molecules. Conversely, nonpolar groups, such as aromatic residues, may try to fold inward upon themselves, creating an island of hydrophobic residues.
(6) They may alter chemical behavior. When weak forces of this kind are established between molecules which approach sufficiently closely, their electronic structures may become deformed, thus altering their chemical reactivity.

Living systems have learned to utilize these forces, which are much weaker than those involved in ordinary chemical bonds, in order to direct the behavior of molecules in highly specific ways, so as to bring about only desired changes in their strong bonds. Thus, instead of relying only on blind random collisions between molecules, living cells begin by orienting the molecules they wish to engage in reaction. This orientation allows only certain molecules to interact and ensures that the reaction produced will be confined to the particular atoms designated within these molecules.

Consider a reaction between two molecules, as the biological system would carry it out. Since the entire reaction occurs at a low temperature, the energy of the random collisions between the two molecules is insufficient to activate them, and, therefore, no spontaneous reaction occurs. However, there is present in the system a large enzyme molecule which may have a molecular weight as high as 100,000 daltons (one dalton is the weight of one hydrogen atom). The enzyme molecule, which is a protein, has a unique three-dimensional configuration possessing regions on its surface specifically shaped to fit molecule A at one point and molecule B at another. The nest that is arranged for each molecule fits it so as to be complementary, both geometrically and in the distribution of electrostatic charges, so that a net attractive force is present. The tailoring is so accurate that the individual atoms of the small molecules and those forming the nest in the surface of the large molecule are brought sufficiently close so that the weak binding forces previously described are brought into play. The net force causing the attachment of the substrate molecule to the enzyme is the resultant of those engendered from about three to five pairs of atoms. While each atomic interaction by itself is weak, in concert they are strong enough to produce attachment that is both highly specific and stable at body temperature. The accuracy of the fit and of its resulting attachment process is so great that virtually no other molecule likely to be present in the cell can become attached at these specific sites designated for the A or B molecule, respectively. Attachment of a specific substrate can induce a change in the enzyme structure so as to lock the substrate molecule into a position in which a specific chemical bond is activated and the desired chemical change occurs. This is accomplished by a process called "orbital steering," in which the proximity of the atoms of the enzyme causes a rearrangement of the electronic motions around the atoms of the substrate in a highly specific fashion. An adjustment in bond angles and bond energies occurs, facilitating a particular mode of chemical transformation in the substrate. Thus, living systems have learned to guide molecules through chemical reactions by using enzymes which open up a single pathway for each molecule. The difference in mode of action between an enzyme and a conventional catalyst is illustrated in highly schematic fashion in Figure I-4.

Virtually all chemical reactions in living cells are mediated by enzymes. Therefore, in any living cell a huge collection of different enzyme molecules is present, each one promoting one specific chemical transformation. But the structure of each enzyme must be specified and accurately controlled. How does the cell manage to produce these molecules over and over again with the degree of specificity and precision known to occur in living systems? The quest for the answers to this question stimulated one of the most exciting periods in the history of science, and the answers obtained, now generally grouped

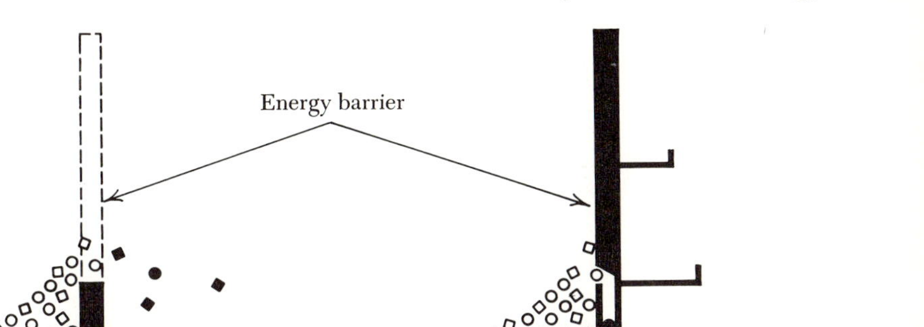

Figure I-4 *Demonstration of the difference in action between enzymes and ordinary chemical catalysts. Both result in a lowering of the energy barrier the molecules must traverse. However, most catalysts permit the molecules that pass the barrier to react in a variety of different ways, while enzymes steer molecules through a single reaction path even when a complex mixture of substances is present. (Redrawn from T. T. Puck, The biological sciences, in "The Great Ideas Today," Encyclopedia Britannica, Inc., Chicago, 1967, with permission.)*

under the heading "molecular biology," laid the foundation for the current revolution in biology.

Review of molecular biology. Molecular biology arose as a conceptual synthesis, resulting from genetic and biochemical experiments in microorganisms. The genetic operations required by these experiments involve (1) the isolation of mutants, a task made easy by the ability to grow specific single cells into colonies whose mutant behavior can readily be recognized, and (2) genetic analysis of the resulting mutants, in which the position of the gene involved and its difference in behavior from the normal gene could be determined.

The biochemical operations consisted in study of the altered metabolic pathways produced by mutation, and isolation and identification of the macromolecules which embody the genetic information.

Structures of DNA, RNA, and protein. That DNA constitutes the gene material was first demonstrated by the work of Avery and his coworkers and by Hershey and Chase, who showed that in the pneumococcus and in *E. coli* bacteriophage, respectively, specific gene activity could be conferred upon cells by transfer of material consisting entirely, or almost entirely, of DNA.

The familiar double-helix structure of DNA, as proposed by Watson and Crick (Figure I-5), was derived from analysis of the x-ray diffraction pattern of DNA obtained by Wilkins and Franklin and by consideration of the known chemical relationships exhibited by DNA, such as the fact that the number of adenine residues was always equal to that of the thymine, and guanine to cytosine. The combination of strong covalent chemical bonds, which hold together the atoms of the individual strands, and the weak forces, which unite the two single strands into the double helix, constitutes a structure unsurpassed for its elegance and efficient informational storage capacity. Thus, one unit of information can be stored in the volume of one nucleotide pair, which is approximately 2×10^{-21} cm^3. The most efficient informational storage systems of modern computers fails to approach this efficiency by many orders of magnitude.

While only one functional kind of DNA is known, a variety of different RNAs exist: messenger, transfer, viral, ribosomal, and, presumably, smaller components of these which are occasionally found in various cells. RNA resembles DNA except that (1) its sugar moiety is ribose instead of deoxyribose; (2) it utilizes the pyrimidine base uracil instead of thymine to pair with adenine; (3) some forms, such as transfer RNA, always contain modified bases, whereas DNA does rarely; and (4) it is almost always single-stranded, in contrast to DNA, which is almost always double stranded. Because of this single-strandedness, the bases of RNA can achieve a double-stranded condition only by pairing with another molecule that might have stretches of complementary base pairs or by doubling the molecule up on itself, so as to bring together base sequences that are complementary as occurs in transfer, ribosomal, and some viral RNAs. Some of the properties of the different kinds of RNA are summarized in Table I-1, and their functional relationships are shown in Figure I-6.

Table I-1 Summary of Properties of Various Types of RNA in Bacteria and Bacteriophage

Type of RNA	*Molecular Weight (daltons)*	*Presence of Modified Bases**	*Approximate Half-Life*	*Function*
Messenger	Highly variable depending on proteins coded (average about 4 x 10^5)	Not reported	1-3 min	Transcribes a copy of the information content of the DNA and translates this information into a particular protein structure by means of the ribosomal machinery
Transfer	2.5 x 10^4	About 12% of all bases	Stable	Attaches to the activated amino acids and transfers them into the growing protein chain, in accordance with the sequence called for by the messenger
Ribosomal	1.8 x 10^6	Few percent	Stable	Constitutes approximately half of the mass of the ribosome, which, together with the messenger, forms the protein-synthesizing machinery
Viral	Variable (in a typical case (F2): 1 x 10^6)	Not reported	Stable	Found in various bacteriophages, as well as animal viruses; combines within itself both the genetic function of DNA and the messenger function of messenger RNA

*All types of RNA contain the usual bases: adenine, guanine, cytosine, and uracil. Some RNAs, however, contain, in addition, certain modified bases: inosine, dimethylguanosine, pseudouridine, dihydrouridine, ribothymidine, methylinosine, and methylguanosine.

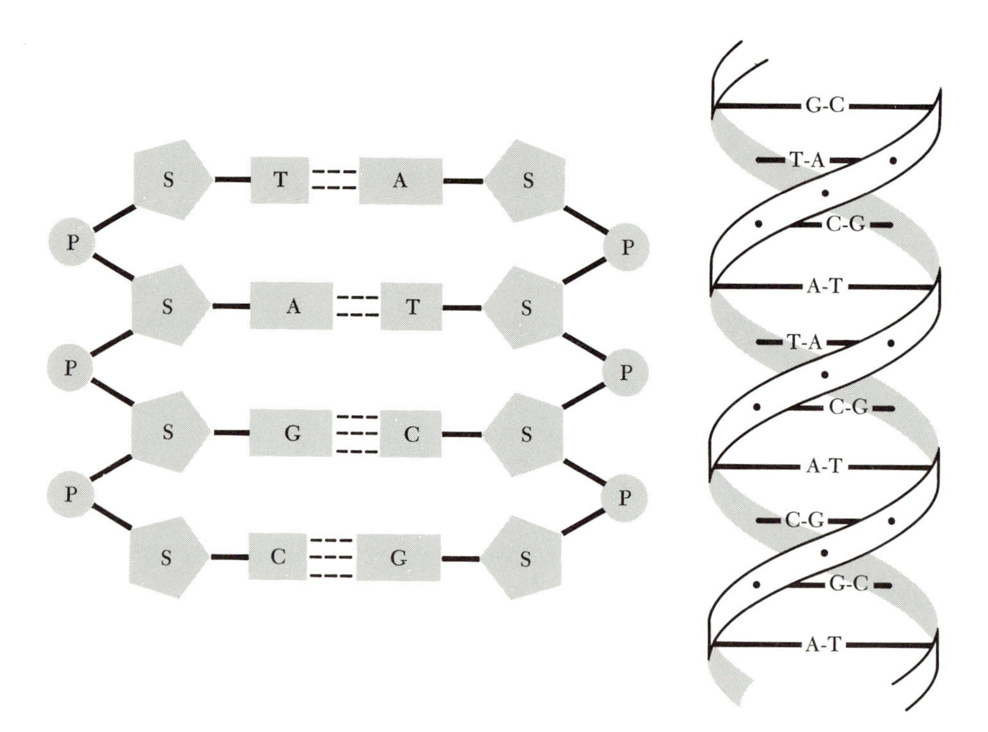

Figure I-5 The structure of DNA, which consists of two single-stranded backbones containing alternating phosphate (P) and sugar (S) entities to which are connected one of the four possible bases, A, T, G, or C. The two strands are held together by hydrogen bonds between these bases in four different arrangements as shown in (a). The molecule might be viewed as a twisted ladder, the supporting bars consisting of the sugar-phosphate backbone, while the rungs of the ladder are made up of the hydrogen-bonded base pairs (b).

Figure I-6 The mechanism by which information in DNA is translated by means of an mRNA blueprint into the amino acid sequence of a protein. During transcription, the original information in the chromosome is encoded in a complementary mRNA. This RNA then forms a complex with the ribosomal subunits, which, in the presence of formyl-methionine tRNA, initiates translation. Growth of the peptide chain can be visualized as the result of a number of concerted reactions. As pictured above, the amino acid valine, attached to its tRNA at the ribosomal binding site A, is shown being incorporated into the growing end of the peptide chain with the concomitant displacement of the tRNA occupying site B. As the mRNA progresses along the ribosome, valine tRNA, to which the growing peptide chain is now attached, is moved to site B thus making available site A for the incoming phenylalanine tRNA and the initiation of a new cycle. The growing amino acid chain proceeds to fold up into its characteristic three-dimensional configuration and is then released from the ribosome when the peptide sequence has been completed.

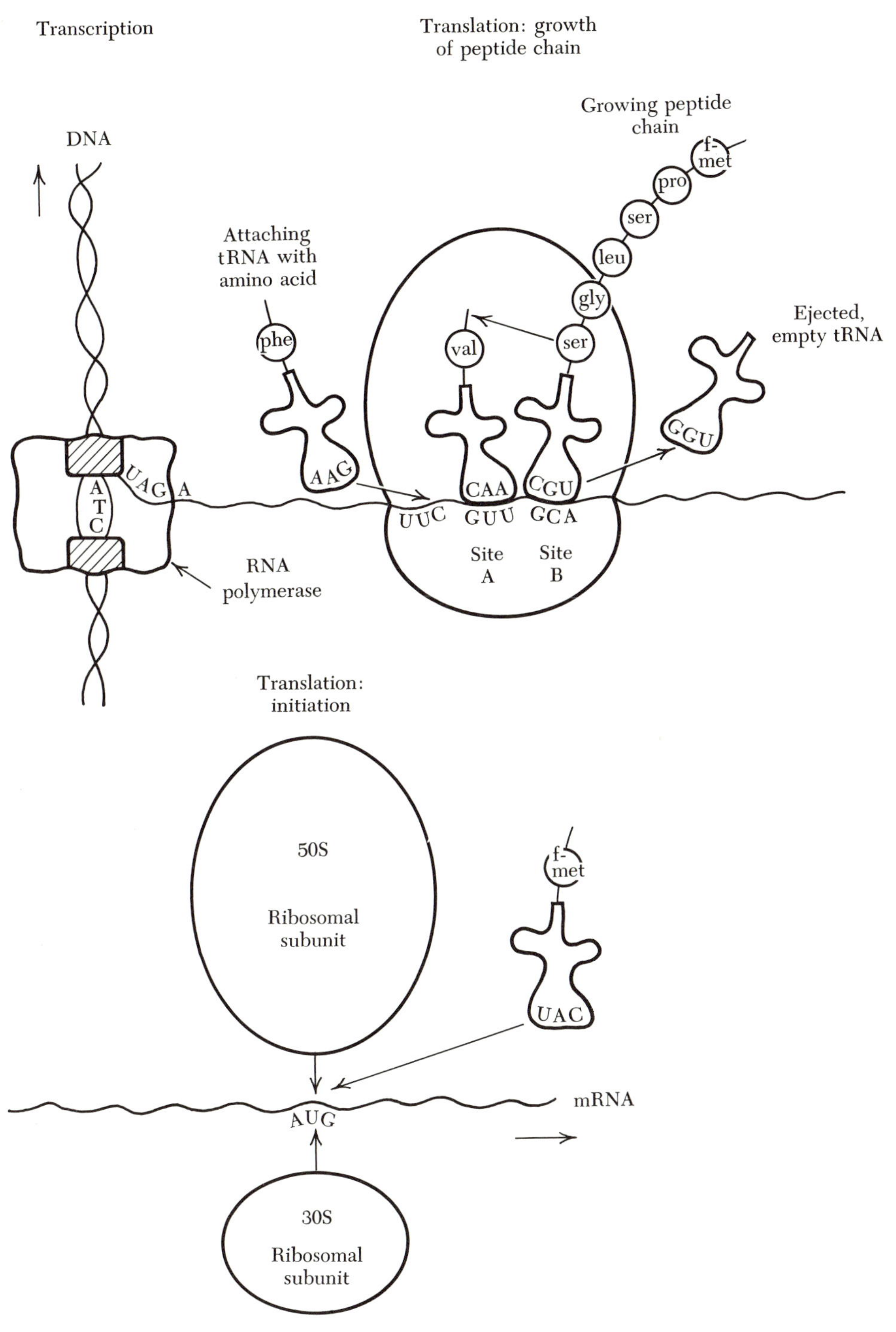
Transcription
Translation: growth of peptide chain
Growing peptide chain
DNA
f-met
pro
ser
leu
gly
Attaching tRNA with amino acid
phe
val
ser
Ejected, empty tRNA
GGU
UAG
A
A
T
C
AAG
CAA
CGU
UUC
GUU
GCA
Site A
Site B
RNA polymerase
Translation: initiation
50S
Ribosomal subunit
f-met
UAC
mRNA
AUG
30S
Ribosomal subunit

X-ray diffraction techniques have been successful in elucidating the three-dimensional structure of proteins as well as of nucleic acids. The protein problem was far more complex because of the lack of the simplicity afforded by the double helical structure. The first protein structures to be elucidated were the mammalian proteins myoglobin and hemoglobin. Since then, others have also been determined, including those of lysozyme, ribonuclease, and chymotrypsin. In large part, the three-dimensional configuration adopted by a protein is determined by the following factors: The amino acid chain tends to form an α-helix, which appears to be its most stable configuration, except where other forces intervene; the introduction of prolines in the chain can serve to distort the α-helical progression; amino acids with strongly hydrophobic groups, such as phenylalanine, have a strong tendency to assume a configuration in which the hydrophobic residues are buried deep within the protein structure, so as to be close to each other and away from the water environment; and, conversely, the highly polar amino acids, such as lysine and serine, tend to project into the surrounding aqueous medium, so as to be closely associated with the water molecules.

X-ray diffraction diagrams of crystalline enzymes, which have attached small molecules whose structures resemble those of the natural substrate, have made it possible to visualize directly the site of attachment of the substrate. In at least one case, it has been possible to observe the structural rearrangement which occurs in the protein as a result of such attachment.

Synthesis of DNA, RNA, and proteins. The mechanism of replication of DNA, which is the heart of the biological reproductive process, was first deduced from its structure, as a separation of the two helical components, followed by enzymatic construction of a new complementary strand along each of the old DNA strands. Such a mechanism made two important predictions. First, it demanded that each double helix contain one strand that has been inherited from the parental DNA and one that has been newly synthesized from small molecules. The elegant experiments of Meselson and Stahl, who worked with *E. coli*, and Taylor who used cells of higher organisms, brilliantly confirmed this prediction (Figure I-7). Second, this mode of DNA formation represents a new principle of chemical synthesis unique to living systems: the synthesis of a complex molecule carried out by a stepwise copying of a linear sequence specified by a preexisting

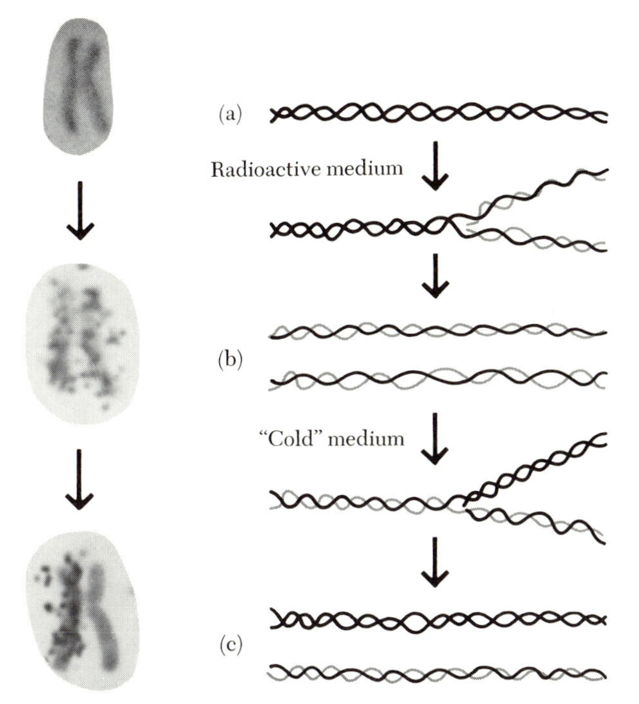

Figure I-7 Experimental verification in plant chromosomes of the Watson-Crick model for DNA replication. (a) Chromosomes dividing before radioactivity has been added. (b) Labeling pattern after one division in radioactive medium. Both strands of each chromosome are uniformly labeled. (c) After one division in radioactive medium followed by one division in "cold" medium, only one of the two strands is labeled, except where crossing-over has occurred. (Photographs courtesy of J. H. Taylor.)

molecule. This mechanism implies that a DNA-synthesizing enzyme could not function unless preexisting DNA could be supplied to act as a template. Kornberg's isolation of this enzyme and the demonstration that it does indeed act in this way was perhaps the crucial experiment that convinced biologists of the essential soundness of the theoretical formulation of molecular biology. In RNA synthesis, the DNA is again the template. In this case, however, RNA is transcribed from only one chain of the DNA, and the enzyme, called RNA polymerase, incorporates into the final structure ribose and uracil instead of deoxyribose and thymine, respectively.

The synthesis of protein follows a more complex course because the linear nucleotide sequence must now be translated into a different language, a linear chain of amino acids linked by peptide bonds. This translation occurs in accordance with the steps outlined in Figure I-6. The gene, which is the ultimate repository for the genetic information specifying the constitution of a specific protein, is accurately transcribed into a messenger-RNA molecule, which is a working blueprint made from the master plan. The information carried by the messenger RNA is then translated into the corresponding linear sequence of amino acids at the ribosome, an exquisitely designed molecular machine that successively takes up, in appropriate order, transfer-RNA molecules, each carrying the amino acid specified by the messenger. The amino acids are successively disengaged from their transfer-RNA moieties and linked together to form the specified sequence.

The predictions of these concepts have been confirmed by the achievement of the *in vitro* synthesis of RNA and protein, as well as of DNA. Such syntheses have verified (1) the need for a DNA template to achieve RNA synthesis; (2) the fact that the RNA formed has a sequential composition of purine and pyrimidine bases complementary to that of the DNA from which it was formed; and (3) the requirement of messenger RNA, ribosomes, transfer RNA, and a set of enzymes and other factors for the synthesis of protein. Once these requirements were defined, the nature of the genetic code, by which base sequences in the DNA are translated into amino acid sequences in the resulting protein, could be attacked. The elucidation of this code, which is used by organisms as diverse as bacteriophage, tobacco mosaic virus, and man, is one of the great intellectual achievements of the age. The structure of the code is shown in Figure I-8.

Control of protein synthesis. A cell which continuously synthesized all of the proteins coded for by all of its genes would be highly inefficient. Initial insights into mechanisms by which protein synthesis is regulated came about in experiments that demonstrated two kinds of enzyme synthesis in bacterial cells. The so-called "constitutive" enzymes are always manufactured by the cell regardless of external conditions. "Induced" enzymes, however, are not synthesized, unless the molecular environment changes, as, for example, when a new medium is substituted, requiring the presence of new enzymes necessary for continued multiplication of the cell. In the study of this phenomenon, it was found that new synthesis is initiated sequentially for all of the enzymes involved in the metabolic chain needed for utilization of a new nutrilite, and that the genes responsi-

First base	Second base: U	C	A	G	Third base
U	UUU, UUC } Phe UUA, UUG } Leu	UCU, UCC, UCA, UCG } Ser	UAU, UAC } Tyr UAA, UAG } Chain terminators	UGU, UGC } Cys UGA Chain terminator UGG Tryp	U C A G
C	CUU, CUC, CUA, CUG } Leu	CCU, CCC, CCA, CCG } Pro	CAU, CAC } His CAA, CAG } GluN	CGU, CGC, CGA, CGG } Arg	U C A G
A	AUU, AUC, AUA } Leu AUG* Met	ACU, ACC, ACA, ACG } Thr	AAU, AAC } AspN AAA, AAG } Lys	AGU, AGC } Ser AGA, AGG } Arg	U C A G
G	GUU, GUC, GUA, GUG } Val	GCU, GCC, GCA, GCG } Ala	GAU, GAC } Asp GAA, GAG } Glu	GGU, GGC, GGA, GGG } Gly	U C A G

Figure I-8 An arrangement first devised by Crick for showing the code by which genetic information is transferred from nucleic acid (in the form of messenger RNA) to protein. Asterisk indicates code involved in chain initiation.

ble for the synthesis of these enzymes are all located adjacent to each other on the chromosome.

As a result of such experiments, the operon theory for the regulation of protein synthesis in bacteria was developed. The scheme is illustrated in Figure I-9. An operon is composed of two controlling genes and a set of structural genes for the enzymes involved. One of the controlling genes is an "operator," which is contiguous to the structural genes and which corresponds to a control switch that can be

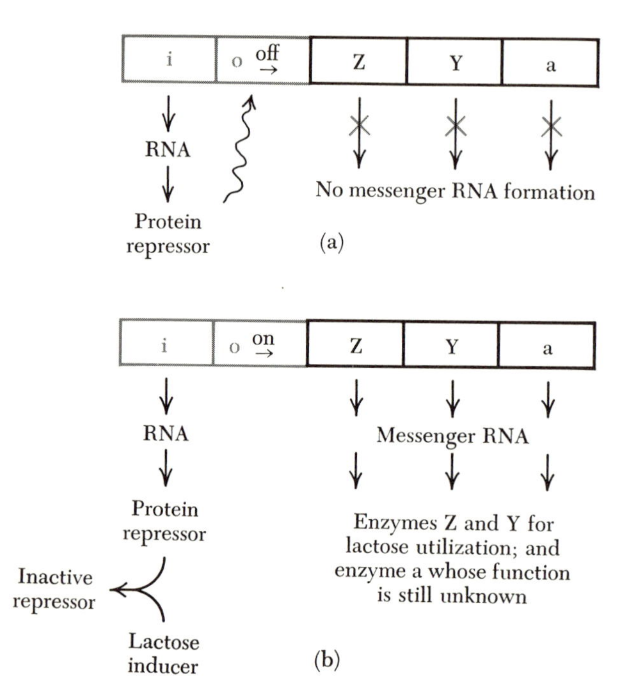

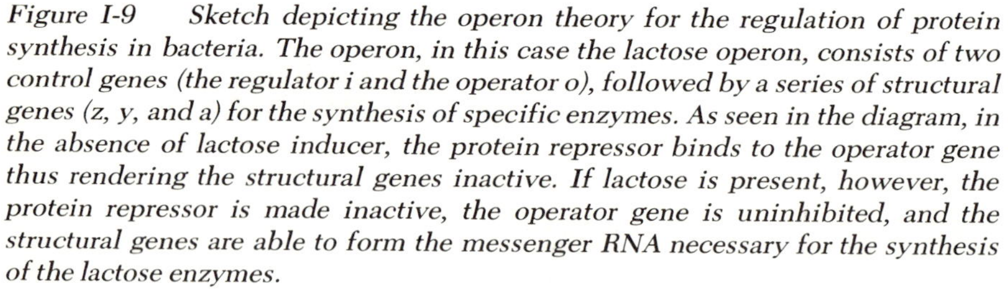
Figure I-9 Sketch depicting the operon theory for the regulation of protein synthesis in bacteria. The operon, in this case the lactose operon, consists of two control genes (the regulator i and the operator o), followed by a series of structural genes (z, y, and a) for the synthesis of specific enzymes. As seen in the diagram, in the absence of lactose inducer, the protein repressor binds to the operator gene thus rendering the structural genes inactive. If lactose is present, however, the protein repressor is made inactive, the operator gene is uninhibited, and the structural genes are able to form the messenger RNA necessary for the synthesis of the lactose enzymes.

set to cause messenger formation to be switched "on" or "off." The setting of the operator to the "on" or "off" position is brought about by a regulator gene. In the example of the lactose operon, which is the one that has been most thoroughly studied, the regulator gene produces an RNA which, in turn, causes formation of a repressor protein. This protein attaches to the operator gene and inactivates the messenger formation from all the other genes of the operon. When the synthesis of an enzyme of a particular operon is needed, specific molecules, called "inducers," are taken up by the cell from the medium and neutralize the action of the repressor. As a result, the operator reverts to the "on" position and the corresponding structural genes become active.

This is an example of negative control, that is, in the absence of inducer, the genes corresponding to the enzymes needed to metabolize it are rendered inactive by the repressor. In order to turn these genes on, the repressor must, in turn, be inactivated by the inducer. In the case of at least one other operon system, that involved in the metabolism of arabinose, there is evidence for an alternative type of control, that is, the production of an activator molecule by the regulator gene that turns on the synthesis of the normally inhibited structural genes of the operon.

Allosteric regulation of enzyme activity. The operon theory illuminated aspects of how enzyme synthesis is regulated. However, this control mechanism can be relatively slow. In addition, then, cells have rapid mechanisms for controlling the activity of enzymes that are already synthesized and exist inside the cell. As discussed earlier, enzymes possess specific sites for the attachment and activation of their substrate molecules. They also contain other specific sites for the attachment of other kinds of molecules, such as the products of the reaction chain in which the enzyme is contained. Thus, when enzyme products are being produced faster than they can be used up, these molecules accumulate and bind to the secondary sites on the enzyme, causing it to undergo a structural rearrangement. This change in the conformation decreases the catalytic efficiency of the enzyme, and the reaction velocity falls until a more balanced situation is restored. One such system is illustrated in Figure I-10.

This phenomenon is called allosterism. Thus, not only do enzymes possess the power to evoke highly specific chemical reactions, but they also contain feedback-control mechanisms, built into their very molecular structures, that can modulate the rate of catalytic action in accordance with the cell's instantaneous needs. The small molecules that attach to a protein and modify its catalytic activity are called effector molecules and may exert either an inhibitory or an activating effect. The binding forces of such effector molecules are of the weak variety (noncovalent) and therefore allow rapid and reversible interaction that can respond rapidly to very slight changes in the cell's chemical environment.

It is interesting to note that, while information for protein structure is coded in a linear fashion in the gene and is expressed as a linear sequence of amino acids in the resulting polypeptide chain, the function of the resulting protein can only be achieved when it has folded

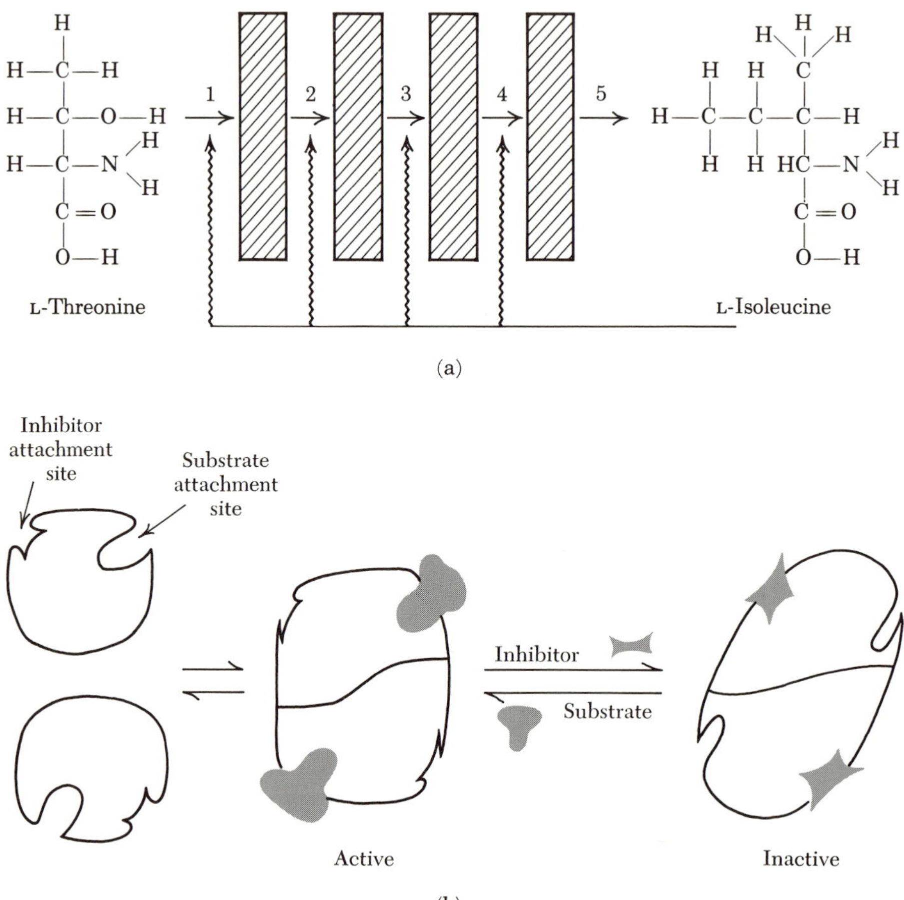

Figure I-10 *(a) Example of a chain of enzymatic conversions starting with L-threonine and proceeding through five steps ending in L-isoleucine. Each of the five enzymes is allosterically inhibited by an excess of the final product, L-isoleucine. (b) A conceptual model of the process of allosteric inhibition of an enzyme. The monomeric forms can unite to yield a polymer (in this case dimer) which can exist in two alternative forms. Equilibrium is driven to the right or left by the excess of inhibitor or substrate molecules respectively.*

up in its characteristic three-dimensional form. The three-dimensional geometry of the protein is strongly influenced by the ionic and molecular environment in which it finds itself, and, in the case of enzymes, the biological activity can be profoundly altered by the nature of the molecular and ionic species that surround and attach to the enzyme. Thus, the linear geometrical arrangement characteristic of the gene is useful for informational storage and for translation into the protein structure, but, thereafter, the function itself requires a more complex three-dimensional configuration.

The complex nature of enzyme action may contain the explanation of why these molecules must be so large. When one reflects that an enzyme requires (1) specific attachment sites on its surface not only for its substrates but also for the necessary molecular arrangements involved in securing the fit of the substrate, (2) the structural parameters for producing activation of the specific bonds, (3) the sites for attachment of allosteric small molecules that regulate the protein's reactivity in given situations, and (4) in the case of some enzymes at least, the sites required for anchoring the enzyme in specific cell structures, it is evident that no molecule with a molecular weight as low as a few thousand could easily embody all these different functions.

Summary. Living processes are carried out inside of each cell by chains of molecular reactions consisting of thousands of highly coordinated individual steps proceeding simultaneously with great precision and speed. Ordinary molecules are stable at room temperature because of the existence of energy barriers, which must be overcome before chemical reaction can occur. Chemical reactivity usually requires heating to high temperatures to surmount these barriers, but, as a result of the randomness inherent in all atomic and molecular events, the resulting reactions rarely proceed to completion, often yielding mixtures of products and exhibiting velocities that are difficult to control accurately.

Living cells appear to be the most highly ordered molecular structures in the universe. They carry out the incredibly varied and exquisitely coordinated chemical processes that underlie the manifestations of life by virtue of their having developed the means to circumvent the randomness inherent in nonliving molecular behavior. This control is achieved by means of enzymes, giant protein molecules that use the weak but specific chemical forces present in every molecule to orient the substrate prior to its reaction, so as to guide the reaction along particular pathways with regulated velocities.

The explosive developments of molecular biology that have occurred in the last decade arose through a combination of genetic, biochemical, and physico-chemical operations. Results from these studies have revealed basic chemical principles operating in bacteria: (1) The fundamental information is stored in each cell in linear, molecular sequences that direct the details of the synthetic reactions producing the cellular machinery; (2) the use of template surfaces allows replication of this information storage system and synthesis of the key macromolecules to be carried out with incredible fidelity; (3) molecular modulational systems exist that control the levels of enzyme activity in accordance with the cell's need at any moment (one such system controls the synthesis of enzymes; another regulates the catalytic efficiency of the enzyme itself).

REFERENCES

Lehninger, Albert L. "Bioenergetics," Benjamin, New York, 1965.

Watson, James D. "Molecular Biology of the Gene," Benjamin, New York, 1965.

APPENDIX II SOME HUMAN CHROMOSOMAL ANOMALIES AND THEIR CHARACTERISTIC SYMPTOMS

Chromosomal Constitution	*Disease*	*Some Characteristic Symptoms*
Sex chromosomes:		
45, X	Turner's syndrome	Failure to complete sexual maturation at puberty; shortness of stature; occasional somatic defects
47, XXY	Klinefelter's syndrome	Male genitals with abnormally small testes; some female sex characteristics such as breast enlargement
47, XXX	Triple X	Female phenotype; sometimes characterized by mental retardation and absence of menses
47, XYY	Extra Y	Greater-than-average tallness; some tendency toward abnormally aggressive behavior; emotional instability
Autosomes:		
47, XY or XX; trisomy 21	Down's syndrome (mongolism)	Mental retardation; multiple somatic abnormalities, including slanting of the eyes
45, XY or XX; translocation between chromosome 21 and another chromosome	Down's syndrome carrier	Balanced translocation: clinically normal but high proportion of offspring with Down's syndrome

Chromosomal Constitution	*Disease*	*Some Characteristic Symptoms*
46, XY or XX; but with a translocation between chromosome 21 and another chromosome	Down's syndrome	Unbalanced translocation: Down's syndrome
46, XY or XX; but enlarged satellite in 13-15 or 21-22	Occurs in about 40% of the cases of Marfan's syndrome; other conditions involving congenital heart defects	Tallness; congenital heart and eye defects
47, XY or XX; trisomy 18	Trisomy 18	Mental retardation; failure to thrive; multiple congenital abnormalities of fingers, ears, mandible, heart, and gastrointestinal tract
47, XY or XX; trisomy 13-15	Trisomy 13-15	Severe multiple abnormalities with absence of olfactory lobe of the brain; cleft palate and harelip; defects of nose and eyes; congenital heart disease; convulsions and mental retardation
46, XY or XX; but bone marrow cells display an abnormally small acrocentric chromosome, probably chromosome 22	Chronic myelogenous leukemia	Chronic leukemia

Chromosomal Constitution	*Disease*	*Some Characteristic Symptoms*
Highly aneuploid karyotype, often with various abnormal chromosome structures	Various kinds of cancer	Malignancy
46, XY or XX; but deletion of the short arm of chromosome 5	"Cri du chat" syndrome	Mental retardation; peculiar "cat" cry; abnormally small head
46, XY or XX; but deletion in long arm of chromosome 13	Chromosome 13 deletion syndrome	Facial asymmetry; small, triangular-shaped head; absent or underdeveloped thumbs; mental retardation
46, XY or XX; but multiple chromosomal breakage	Bloom's syndrome	Varied symptomatology, including frequent cancer, offspring with chromosomal deletions and aberrations, and blood disorders

APPENDIX III CALCULATION OF EXPECTED FREQUENCY OF SPONTANEOUS ABORTION FROM CHROMOSOMAL ANOMALIES

Assume that the incidence of trisomy 21 is representative of nondisjunctional processes, in which only the embryo with the extra chromosome is viable. The incidence of this trisomy is about 0.0025 and is similar to that of the XXY trisomy. Assume that this frequency will be similar for each chromosome and will be equal to that for the corresponding monosomy as well.

Only four chromosome pairs occur in the trisomic condition in man with a frequency at all near the figure given above. Hence, a trisomy in any of the remaining 19 chromosome pairs may be taken as almost always lethal, so that the probability of a lethal trisomy would be 0.0475. Similarly the only monosomy which occurs in appreciable numbers in man is that of Turner's syndrome, XO. Let us assume that a monosomy in any of the 22 autosome pairs is almost always lethal. Hence the probability of a lethal monosomy is approximately 0.055, and the probability of a lethal trisomy or monosomy in any human fertilization process would be about 10%. While this figure is admittedly approximate, it would appear to establish at least an order of magnitude for the frequency of abortive events to be expected, as a result of chromosomal anomalies in the original zygote, and agrees reasonably well with that estimated by gynecologic experience.

It may well be that the frequency of nondisjunction is not completely random for large and small chromosomes, since the large chromosomes might have a greater tendency to become entangled in microtubular spindle fibers and so be involved more often in nondisjunction. There is also reason to expect that the larger chromosomes would have a frequency roughly proportional to their size for undergoing chromosomal breakage.

APPENDIX IV REPRESENTATIVE SINGLE-GENE MUTATIONS

Condition	Protein Deficiency or Abnormality
Carbohydrate metabolism:	
Hepatorenal glycogenosis	Glucose-6-phosphatase
Galactosemia	Galactose-1-phosphate uridyl transferase
Amino acid metabolism:	
Phenylketonuria	Phenylalanine hydroxylase
Alcaptonuria	Homogentisic acid oxidase
Lipid metabolism:	
Acanthocytosis	β-lipoprotein
Gaucher's disease	Glucocerebroside-cleaving enzyme
Purine or pyrimidine metabolism:	
Xanthinuria	Xanthine oxidase
Orotic aciduria	Orotidylic acid pyrophosphorylase and decarboxylase
Lesch-Nyhan syndrome*	Hypoxanthine-guanine pyrophosphoribosyl transferase
Porphyrin and heme-pigment metabolism:	
Acute intermittent porphyria	Increase of δ-aminolevulinic acid synthetase
Familial nonhemolytic anemia	Glucuronyl transferase
Erythrocyte protein:	
Favism*; drug-induced hemolytic anemia	Glucose-6-phosphate dehydrogenase
Congenital nonspherocytic anemia, type II	Pyruvate kinase
Sickle cell anemia (and other hemoglobinopathies)	Hemoglobin (specific amino acid substitutions)
Plasma proteins:	
Agammaglobulinemia*	Gamma globulin
Wilson's disease	Ceruloplasmin
Clotting mechanisms:	
Hemophilia A*	Antihemophilic globulin
Hemophilia B*	Plasma thromboplastin component
Pigment protein:	
Color blindness*	Specific retinal pigment proteins

*Sex-linked

APPENDIX V THE ACCUMULATION FUNCTIONS FOR MAMMALIAN CELLS BLOCKED IN MITOSIS

Consider a randomized cell culture growing with a mean doubling time T. If the variance of T is small, the number of cells, n, present at any moment is

$$n = n^0 2^{t/T} \tag{1}$$

where n^0 is the cell number at $t = 0$. Let N_1 be the fraction of the population contained within the interval of the life cycle, T_1, defined as the period extending from the end of mitosis to a given point 1, and measured counterclockwise around the cycle, as shown in Figure V-1. Then, for the four intervals, G_1, S, G_2, and M (mitosis)

$$T_{G_1} + T_S + T_{G_2} + T_M = T \tag{2}$$

$$N_{G_1} + N_S + N_{G_2} + N_M = 1 \tag{3}$$

Obviously, each of these divisions can be subdivided further if desired. The fraction of the cell population contained in the interval T_1 of Figure V-1 is constant for a random culture and has been shown by other investigators (Crick, 1952; Stanner and Till, 1960; Smith and Dendy, 1962) to be[1]

$$N_1 = 2^{T_1/T} - 1 \tag{4}$$

This well-known equation makes it clear that the cell population is not distributed uniformly over the period of the life cycle, but is most

[1]At $t = 0$, the cell number is n^0, and the fraction of cells in T_1 is N_1. If a time $t = T_1$ is now allowed to elapse, all the cells in T_1 will double, and the new cell population is $n = n^0 2^{T_1/T}$. The increase in cell number, Δn, is equal to those originally present in T_1 and is $n^0(2^{T_1/T} - 1)$. Hence, $N_1 = \Delta n/n^0 = 2^{T_1/T} - 1$.

dense immediately after mitosis and falls smoothly to a minimum just before cell division.

Consider the change in distribution that results on addition of a blocking agent which acts only at a specific point in the cycle. It is assumed that the cells contained in a given interval preceding this point can be recognized by a visual test.

Case I. The blocking agent acts instantaneously and only at the end of mitosis (i.e., at $T = 0$), so that cell multiplication stops when the agent is added. In this case, the total cell number n is constant:

$$n = n^0 = \text{constant} \tag{5}$$

Since mitosis itself is the period immediately preceding the block and can be readily scored visually, the accumulation of cells in mitosis is measured. If the blocking agent is added at $t = 0$, and the proportion of cells in mitosis, N_M, scored at different intervals, the relationship will be obtained in which N_1 is N_M:

$$N_1 = 2^{(T_1 + t)/T} - 1 \tag{6}$$

This situation differs from that of Equation (4) because, as time progresses, cells are prevented from leaving the region T_1 (Figure V-1). Hence, the effect is as though the upper limit of T_1 is increasing with time so as to sweep out a progressively larger region. By rearrangement, Equation (6) becomes

$$\log_{10} (1 + N_1) = \frac{0.301}{T} (T_1 + t) \tag{7}$$

We call the quantity $\log (1 + N)$ the collection function. Thus, for an inhibitor acting at $t = 0$, a plot of $\log (1 + N_M)$ against time t would yield a straight line with intercept equal to $0.301 T_1/T$, where the interval T_1 is T_M, the time of mitosis. Given the type of inhibitor postulated, the straight-line behavior establishes that the cell population was indeed random at $t = 0$ and provides values for the time of mitosis, T_1, and the total generation time, T, from the intercept and the slope.

Case II. Next, consider an agent that instantaneously and quantitatively produces a block at a point, T_1, in the life cycle which precedes mitosis (Figure V-2). The cells in the interval $T_2 = T_1 = \Delta T$ are presumed to be scorable microscopically. The cell fraction present in

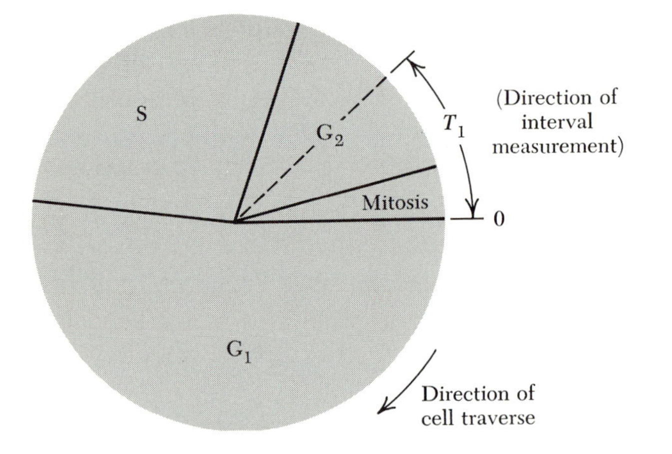

Figure V-1 Diagrammatic representation of the four phases of the mammalian cell life cycle. While the cells traverse this cycle in a clockwise direction, the convention adopted here counts time in a counterclockwise direction beginning from the end of mitosis which is labeled 0. Thus, the arbitrarily chosen interval T_1 represents the fraction of the life cycle shown, and N_1 is the fraction of the total cell population contained within this interval, which would in this case extend from the end of mitosis to the middle of G_2.

the interval ΔT in a random, noninhibited population is given from Equation (4) by

$$\begin{aligned} N_2 - N_1 &= N_{\Delta T} \\ &= 2^{T_2/T} - 1 - (2^{T_1/T} - 1) \\ &= 2^{T_2/T} - 2^{T_1/T} \\ &= 2^{T_1/T}\,(2^{(T_2 - T_1)/T} - 1) \end{aligned} \qquad (8)$$

If the agent is added at $t = 0$, the cells continue to enter the region ΔT, but cannot leave it and so accumulate. However, the cells initially in T_1, being unaffected, continued uninterruptedly around the cycle, so that T_1 is gradually depleted. Moreover, all of these latter cells will divide, so that the cell number does not remain constant, as in Case I, but increases over the period $T_1 \geq t \geq 0$ to reach a constant limit at $t = T_1$:

$$\begin{aligned} n &= n^0 2^{t/T} && \text{for } T_1 \geq t \geq 0 \\ n &= n^0 2^{T_1/T} && \text{for } t \geq T_1 \end{aligned} \qquad (9)$$

The expression for the fraction of the cell population in ΔT at any time after addition of the inhibitor can be found by separately determining the cell fraction of T_2 and T_1 and then subtracting the latter from the former.

If the cell block had been at $T = 0$ in Figure V-2, the fraction of the cell population in the entire region T_2 would be

$$\frac{n_{T_2}}{n^0} = 2^{(T_2 + t)/T} - 1$$

Since, however, the block is at T_1, the actual value will be decreased because of the passage of the cells out of T_1. Let us first consider only the time $t \geq T_1$.

$$\frac{n_{T_2}}{n^0} = (2^{(t_2 + t)/T} - 1) - (\text{cell loss from } T_1)$$

Since cells continue to leave T_1 but no new cells enter,

$$\frac{n_{T_1}}{n^0} = 2^{T_1/T} - 2^{t/T} \tag{10}$$

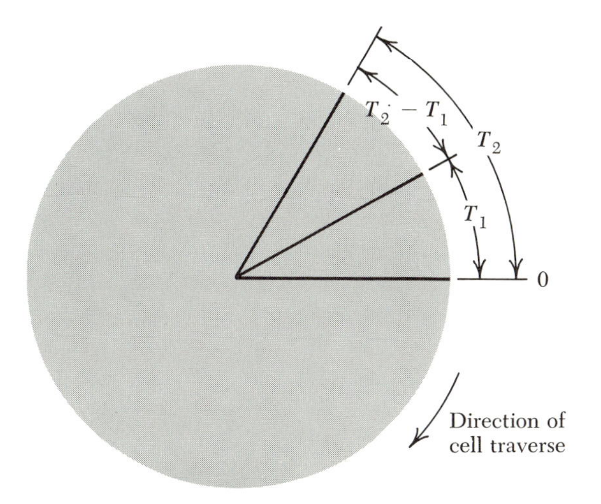

Figure V-2 Life-cycle diagram for a culture in which the cell population of the interval T_2 - T_1, which extends from the end of T_1 to the end of T_2, is to be determined. The fraction of the population of a random culture contained in the interval T_2 - T_1 is given by Equation (8).

Therefore, at any time t, the cell loss from T_1 is 1. Therefore,

$$\frac{n_{T_2}}{n^0} = 2^{t/T}\,(2^{T_2/T} - 1) \tag{11}$$

Now, the cell number in the interval $T_2 - T_1 = \Delta T$ is found by subtracting Equations (10) from (11):

$$\frac{n_{\Delta T}}{n^0} = \frac{n_{T_2}}{n^0} - \frac{n_{T_1}}{n^0} = 2^{(T_2+t)/T} - 2^{T_1/T} \tag{12}$$

Since the cell number is increasing by virtue of the doubling of the cells passing out of T_1, the observed fraction of the cells in the interval ΔT must be corrected by the factor, $1/2^{t/T}$. Therefore

$$\begin{aligned} N_{\Delta T} &= N_2 - N_1 \\ &= \frac{1}{2^{t/T}}\,(2^{(T_2+t)/T} - 2^{T_1/T}) \\ &= 2^{(T_1-t)/T}\,(2^{(T_2-T_1+t)/T} - 1) \qquad \text{for } T_1 \geq t \geq 0 \end{aligned} \tag{13a}$$

$$\begin{aligned} N_{\Delta T} &= \frac{1}{2^{T_1/T}}\,(2^{(T_2+t)/T} - 2^{T_1/T}) \\ &= 2^{(T_2-T_1+t)/T} - 1 \qquad \text{for } t \geq T_1 \end{aligned} \tag{13b}$$

To obtain appropriate linear functions, Equations (13a) and (13b) can be transformed as follows:

$$\log\left(1 + \frac{N_{\Delta T}}{2^{(T_1-t)/T}}\right) = \frac{0.301}{T}\,(T_2 - T_1 + t) \qquad \text{for } T_1 \geq t \geq 0 \tag{14a}$$

$$\log\,(1 + N_{\Delta T}) = \frac{0.301}{T}\,(T_2 - T_1 + t) \qquad \text{for } t \geq T_1 \tag{14b}$$

REFERENCE

Puck, T. T., and J. Steffen. Life cycle analysis of mammalian cells. I. A method for localizing metabolic events within the life cycle, and its application to the action of colcemide and sublethal doses of X-irradiation, *Biophys. J.* 3, 379 (1963).

GLOSSARY

Acrocentric A term denoting a chromosome whose centromere is in a near-terminal position, so that one chromosome arm is much longer than the other.

Aneuploid A chromosomal complement with extra or missing chromosomes.

Autosome Any chromosome other than the sex chromosomes.

Auxotrophic cell A nutritionally deficient cell whose growth requires specific supplementation of the normal growth medium which is adequate for the wild-type strain.

Centromere The region in a mitotic or meiotic chromosome to which the spindle fibers attach.

Chromatid One of the two identical, visibly distinct members of the doubled mitotic chromosome.

Chromosome The structural element of the genetic material on which the genes are contained in a linear sequence. The number and structure of the chromosomes is species specific. In cells of higher organisms, these structures are complex and highly organized, containing RNA and large amounts of protein.

Complementation A process by which two genomes are brought together in a single cell, in such a way that a deficiency in one may be restored or "complemented" by the other.

Complementation analysis Determination of whether cells with the same phenotype have the same genotype. For example, two glycine-deficient mutants are fused, and the hybrid examined for a glycine requirement. If the hybrid no longer requires glycine for growth, the two parental cells were defective in different genes.

Differentiation All the necessary changes in cell structure and function that occur in the course of the normal processes leading from the fertilized egg to the adult individual. These include cell migration, cell multiplication, specific gene and chromosome activation or repression, formation and destruction of specific macromolecules or macromolecular aggregates, and activation or inactivation of specific enzymes.

Diploid The normal state of somatic cells of higher organisms, in which each chromosome, except the sex chromosomes, is represented twice.

Dominant A genetic trait that is expressed either in the heterozygous or homozygous state.

Epithelial In tissue culture, this term applies to cells whose morphology is compact and resembles that of squamus epithelium as opposed to spindle-shaped, fibroblastlike cells.

Euploid The condition in which all the chromosomes are present in constant, whole multiples of the haploid state.

Fibroblast An elongated, spindle-shaped cell which manufactures collagen, giving rise in the body to connective tissue.

Gel electrophoresis A method for separating proteins according to their charge and size. A mixture of proteins is placed on the surface of the gel and subjected to an electric field. The proteins will migrate at specific rates depending upon their net charge and ease of movement through the gel matrix.

Generation time The time required by a population to double.

Genetic carrier An individual heterozygous for a defective recessive gene. Such a person may be clinically normal, but may participate in matings that produce offspring with the homozygous disease condition.

Genetic marker An indicator, usually a mutant gene or chromosome, used to trace the distribution of hereditary traits among offspring of particular matings.

Haploid A chromosomal complement in which each chromosome is represented only once.

Hemizygous A condition in which a normally diploid gene sequence is present in the haploid condition, as in a cell with a chromosomal monosomy.

Heterokaryon A cell containing two or more nuclei that originated from separate cells.

Heterozygous The condition in which specific genetic loci on homologous chromosomes are not identical.

Homologous chromosomes Members of the same chromosome pair. These possess the same genetic loci.

Homozygous The condition in which specific genetic loci are identical on homologous chromosomes.

Hyperdiploid A chromosome complement that contains more than the normal diploid number of chromosomes.

Hypodiploid A genetic constitution that contains less than the normal diploid number of chromosomes but more than a haploid set.

Information That which increases the predictability of a specific outcome for a system.

Interphase All parts of the cellular life cycle other than mitosis (or meiosis).

In vitro A term applied to studies under artificially controlled conditions outside the body.

In vivo A term applied to studies carried out in the intact animal.

Karyotype Arrangement of the chromosomal complement of an individual with homologous chromosomes paired, and the order of the pairs following a prescribed arrangement.

Meiosis The process by which the replicated chromosomes of a sex cell are reduced from the tetraploid to the haploid state by two successive divisions.

Metacentric A term denoting a chromosome whose centromere is located in a position such that the lengths of the two chromatid arms are approximately equal.

Mitosis That portion of the cellular life cycle of higher organisms in which the chromosomes, which have been duplicated in earlier stages of the life cycle, are equally distributed among the two daughter cells.

Monosomy The condition in which only one member of a chromosome pair is present. When only a portion of a chromosome is involved, it is said to be a partial monosomy.

Mutation Any change in the structure of the genome other than those resulting from sexual recombination. Single-gene mutations involve changes in the DNA sequence of a specific gene; chromosomal mutations involve microscopically identifiable changes in the number or structure of any of the chromosomes.

Operon A region of bacterial DNA containing a series of structural genes that are transcribed as a unit, and several regulatory genes that control their transcription.

Polysomic A condition in which one or more chromosomes of an otherwise diploid cell are represented more than twice.

Prototrophic cell A cell having the same nutritional requirements as the wild-type strain.

Recessive A genetic trait expressed only in the homozygous state. However, expression of a genetic trait may depend on the means employed for its identification. Thus, the sickling trait in man is expressed as the disease, sicklemia, in the homozygous state. While heterozygous persons do not display this disease, they will express a difference from normal homozygous persons when their lysed red blood cells are analyzed electrophoretically.

Revertant A mutant which has regained, either partially or completely, the behavior of the wild type.

Somatic A term referring to cells of an organism other than the sex cells.

Time-lapse photomicrography The process by which photographs are made in a microscope at repeated intervals of time and then displayed on a screen so as to simulate a continuously changing situation.

Tissue culture Growth and maintenance of cells outside the body where they can be studied under controlled conditions.

Trisomy The condition in which one or more chromosomes is present in three copies.

Uncertainty principle The existence of a fundamental randomness in nature such that the more precisely the position of a particle is known, the less accurately its velocity can be specified, and vice versa. This principle follows from quantum-mechanical considerations, and the degree of uncertainty involved becomes larger, the smaller the particles which are considered.

Van der Waals bond A weak force engendered between two atoms or molecules through the induction of complementary dipoles in each other by the action of their mutual electrical fields.

Wild type The naturally occurring form of a given species often taken as the prototype for genetic experiments.

INDEX